James A. Saxon, Ph.D.

BASIC
DATA PROCESSING
MATHEMATICS

PRENTICE-HALL, INC.
Englewood Cliffs, New Jersey

ISBN: 0–13–058909–8

Library of Congress Catalog Card Number: 73–171846

10 9 8 7 6 5 4 3 2 1

Printed in the United States of America

PRENTICE-HALL INTERNATIONAL, INC., *London*
PRENTICE-HALL OF AUSTRALIA, PTY. LTD., *Sydney*
PRENTICE-HALL OF CANADA, LTD., *Toronto*
PRENTICE-HALL OF INDIA PRIVATE LIMITED, *New Delhi*
PRENTICE-HALL OF JAPAN, INC., *Tokyo*

Everyone should firmly persuade himself that none of the sciences, however abstruse, is to be deduced from lofty and obscure matters, but that they all proceed only from what is easy and more readily understood.

DESCARTES
(French philosopher and mathematician; 1596–1650)

The function of a teacher (or author of teaching books) is to transform the difficult topic into simple, easy to understand, units of instruction. He should not concern himself with methods and terminology aimed solely at enhancing his reputation for erudite expression. It is hoped that this text lives up to these principles.

THE AUTHOR

CONTENTS

2 NUMBER SYSTEMS   17

3 REVIEW OF ALGEBRA

4 BOOLEAN ALGEBRA

5 INTRODUCTORY LOGIC   99

6 FUNCTIONS AND EQUATIONS   124

7   MATRICES   159

PREFACE

Every discipline has its own terminology and symbology. When you first study a subject such as statistics, you are soon wading through terms such as *mean, median, mode, central tendency* and *correlation coefficient.*

Mathematics is no exception to this rule. Each different area has its own peculiar names, definitions and symbols. Many of these symbols have been developed as a means toward simplification of expressions and ideas. Unfortunately for the beginner, the wealth of new symbology makes the study of the topic harder than it really is until the symbols and their meanings have been mastered.

In this text, a deliberate attempt has been made to simplify the problem by grouping symbols and definitions into tables that can be used as a constant point of reference for the beginner. Appendix D contains a complete listing of all symbols and terminology introduced in the text. A few terms were deliberately omitted since it was felt that they added nothing to the knowledge of computer mathematics, although they would be quite pertinent to a study of pure mathematics. The necessary equations are explained in simple language, with enough examples and problems to make the material understandable to non-mathematically oriented people. Mathematical terminology is deliberately avoided wherever possible.

This text has been developed to be used in a single semester junior college, college or university course in Computer Mathematics, particularly in the field of business. The author has used all of the material in teaching at the adult level for a considerable period of time. Often, students had no formal background in mathematics, but were able to assimilate the material without too much effort because of the nontechnical approach to the topics and the large number of examples and problems in each chapter.

Considerable thanks must be expressed to Jane Wittkow and Jay Caldwell, who provided comments and constructive criticism of the material and to Marilyn Scott, who did her usual admirable job of typing the manuscript.

JAMES A. SAXON

San Diego, California

1 SETS
AND SET THEORY

INTRODUCTION

A *set* is a group or collection of entities that can be specifically defined. In other words, the individual items can be identified as either being a part of a group of items, or not being a part of the group. If the individual items cannot be identified in this manner, the items are not part of a mathematical set.

EXAMPLES

The letters A through G
The numbers 6 through 14
The names TOM, DICK, HARRY
The names of the 50 states
(Note that each item of each set can be specifically identified.)
All the people on your block who are free thinkers
The children in a classroom that will grow up to be politicians
(These last two are not mathematical sets because each item within a set cannot be identified.)

A set may be *finite* (having definite limits) or *infinite*, when referring to the *elements* making up the set. The letters of the alphabet may be called a *finite set* (exactly 26 elements), while the set of counting numbers may be called an *infinite set* because it has no end (1 2 3 4 . . . ; the three dots indicating continuation into infinity).

A set can be given any name when it is being defined. For example, the set consisting of elements TOM, DICK, HARRY, may be called set S.

$$S = \text{TOM, DICK, HARRY}$$

With this definition, we can say with absolute certainty that TOM is an element of set S and JOE is an element *not* in set S.

Often, braces { } are used on either side of a list of elements in a set. In this case, each element within the set is separated from the next element by a comma.

$$H = \{a, b, c, d\}$$

The above statement says that H is a set, consisting of the elements a, b, c, and d. The sequence of the elements within a set is unimportant.

$$H = \{a, b, c, d\} \qquad I = \{d, b, a, c\}$$

The two sets are identical since they contain exactly the same elements.

A set may also contain no elements. This would be called the *null* or *empty* set. In this case, there would be nothing between the braces. This can be written in one of two ways:

$$J = \{ \quad \} \quad \text{or} \quad J = \varnothing$$

One convention often used is to name sets with capital letters and elements within sets with lower-case letters.

EXAMPLE

$$B = \{a, j, k, m\}$$

If you wished to say that m is an element of set B (m belongs to set B), a special symbol, $\in$, may be used.

This is merely a time saver, since it is faster to write "$m \in B$" than "m is an element of set B." To say that an element is *not* a member of a particular set, simply slash the symbol, $\notin$. Thus, $a \notin B$ says "a is not an element of set B."

REVIEW QUESTIONS

1.1. Set B includes the first, third, and fifth day of the week. Is set B finite or infinite?

1.2. Write the following sets.
   (a) Set name F, elements b, c, d, e
   (b) Set name HELLO, elements NOW, HERE, THEN
   (c) Set name J, elements even counting numbers between 5 and 9
   (d) Set name K, elements all days of the week starting with the letter S
   (e) Set name L, elements all days of the week starting with the letter A

1.3. Show, with symbols, that:
   (a) c is an element of set F
   (b) 6 is an element of set J
   (c) Monday is not an element of set K

SUBSETS

If two sets are related in such a way that every element of one set is also an element of the other set, the first set may be called a *subset* of the second set (you may say that one set is *contained* within the other set).

EXAMPLES

   1. $B = \{a, b, c, d, e\}$
      $C = \{a, b, c\}$

Therefore, C is a subset of B. This may be shown with a new symbol, $\subset$, meaning "is a subset of." $C \subset B$.

   2. If all fourth graders in a particular school are in set G and all children in the entire school are in set P, then $G \subset P$ (G is a subset of P).

   3. The western states are in set M and the western and midwestern states are in set N. $M \subset N$ (M is a subset of N).

Two additional facts become apparent from this discussion: (1) every set is subset of itself and (2) the null set is a subset of every set.

Universal Set

Any discussion of sets must be placed in reference to some total or *universal* set, within which the particular set exists or functions. In example 2 above, the universal set would be the entire school population. In example 3, the universal set would be all 50 states, both M and N being subsets of the universal set. When talking about the universal set, we will designate it with the capital letter U.

Pictorial Representation of Sets

A simple way to show sets and their relationships to each other is by means of schematic drawings, called *Venn diagrams*. In these diagrams, a rectangle usually represents the universal set, within which are circles to represent sets.

EXAMPLES

1. $C = \{a, b, c, d\}$ $D = \{e, f, g, h\}$

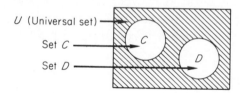

The universal set here is the entire alphabet.
Set C forms one portion of the universe and set D another portion.

These are called *disjoint* sets, where both sets belong to the same universe, but have no elements in common.

2. $E = \{1, 2, 3, 4, 5\}$ $F = \{1, 3, 5\}$

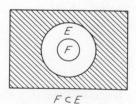

$F \subset E$

Notice here that set F is totally within set E, which shows pictorially that F is a subset of E.
The universal set here is all counting numbers—an infinite set.

REVIEW QUESTIONS

1.4. Interpret the meaning of the following.
(a) $C \subset C$
(b) $\varnothing \subset \{a, b, c, d\}$

1.5. Draw Venn diagrams to show the relationships indicated below.
(a) $A = \{$first six months of year$\}$
$B = \{$second quarter of year$\}$
(b) $C = \{$all even numbers under 10$\}$
$D = \{$all odd numbers under 10$\}$

1.6. What is the meaning of the following?
(a) $d \in k$
(b) $f \notin L$

1.7. Show the universal set of question 1.5(b).

1.8. Which of the Venn diagrams shown below indicate that C is a subset of B and A is a subset of B?

(a)

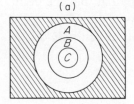

(b)

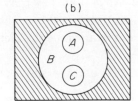

(c)

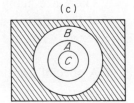

1.9. Using mathematical notation, indicate the relationships in the three Venn diagrams above.

TYPES OF SETS

Identical Sets

When two sets contain exactly the same elements (in whatever order), they are *identical*. In this case, an equal sign may be used, but remember that the meaning of "equal" in this event is that the two sets contain the same elements.

EXAMPLE

$A = B$ (A contains the same elements as B)
$A = \{\text{TOM, DICK, HARRY}\}$ $B = \{\text{HARRY, DICK, TOM}\}$

Equivalent Sets

Sets that are identical are also *equivalent*. Equivalent means that there are as many elements in one set as there are in the other set. Two sets may be equivalent without being identical. The one-to-one correspondence is the basis for determining equivalence.

EXAMPLE

$$A = \{1, 3, 5, 7\}   B = \{2, 4, 6, 8\}$$

These two sets can be paired on a one-to-one basis.

$$A = \{1, 3, 5, 7\}$$
$$\uparrow \ \uparrow \ \uparrow \ \uparrow$$
$$\downarrow \ \downarrow \ \downarrow \ \downarrow$$
$$B = \{2, 4, 6, 8\}$$

There are many other ways that they may be paired, but the point is that the one-to-one relationship is there; therefore, the two sets are equivalent. They are not identical, since the elements are entirely different.

A special symbol may be used to show equivalence:

EXAMPLE

$A \longleftrightarrow B$ (set A is equivalent to set B)

Disjoint Sets

Sets that have no elements in common are called *disjoint* sets. As we have seen in the preceding discussion, disjoint sets may be equivalent, but they can never be identical.

EXAMPLES

$A = \{x, y, z\}$ $B = \{a, b, c\}$ $A \longleftrightarrow B$
$G = \{1, 3, 5, 7\}$ $H = \{9, 11, 13\}$

Nonidentical Sets

In the example above, sets G and H are examples of two sets that are not identical and are not equivalent. Disjoint sets are nonidentical, but they may be equivalent. A set that is a subset of another set is a third type of nonidentical set.

EXAMPLES

1. $A = \{1, 3, 5\}$ $B = \{8, 10, 12\}$ disjoint equivalent
2. $C = \{7, 9, 11\}$ $D = \{7, 9, 19, 21\}$ some elements in common
3. $E = \{13, 15, 17\}$ $F = \{13, 17\}$ F is a subset of E
   $(F \subset E)$

These relationships can be pictorially shown by Venn diagrams:

(1)   (2)   (3)

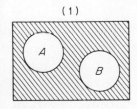 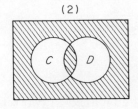 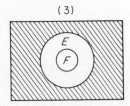

Now consider the following three sets:

$$A = \{a, g, p, r\}$$
$$B = \{g, s, j, p\}$$
$$C = \{p, r, s, t\}$$

A single Venn diagram can show the interrelationships of these sets.

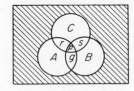

A and B have elements
g and p in common
B and C have elements
p and s in common
A and C have elements
p and r in common

Element p is common to all three sets.

REVIEW QUESTIONS

1.10. How many ways can the elements in the following two sets be paired? This has not been discussed, but try to figure it out.

$$A = \{1, 3, 5\} \qquad B = \{2, 4, 6\}$$

1.11. Draw a Venn diagram showing the relationship of the two sets in problem 1.10. Explain the relationship.

1.12. Draw Venn diagrams showing the relationship of each of the following groups of sets. Describe each group with mathematical symbols, if possible.
(a) $J = \{$Pete, Joe, Jack, Harry$\}$ $K = \{$Pete, Harry$\}$ $L = \{$Pete$\}$
(b) $M = \{1, 3, 5\}$ $N = \{5, 3, 1\}$
(c) $O = \{$all weekdays$\}$ $P = \{$weekend days$\}$
   $Q = \{$Sunday, Monday, April$\}$

WORKING WITH SETS

Union of Sets

When every element contained in two sets is contained in a third set, the third set is called a *union* of the other two sets.

EXAMPLES

$$A = \{p, g, r, s\} \qquad B = \{a, b, c\}$$
$$D = \{a, b, c, p, g, r, s\}$$

Set D is the union of sets A and B. The symbol for union is $\cup$.

EXAMPLE

$$\{p, g, r, s\} \cup \{a, b, c\} = \{p, g, r, s, a, b, c\}$$

or

$$A \cup B = \{p, g, r, s, a, b, c\}$$

or

$$A \cup B = D \text{ (the union of } A \text{ and } B \text{ is } D)$$

If some of the elements in two or more sets are common to both sets, they need only be listed once in the union.

EXAMPLE

$$A = \{1, 3, 5, 6\} \qquad B = \{2, 3, 4, 6\}$$
$$A \cup B = \{1, 2, 3, 4, 5, 6\}$$

Although elements 3 and 6 are present in both sets A and B, they are listed only once in the union of the two sets.

The union of two sets can be shown clearly in a Venn diagram by shading the union of the sets.

EXAMPLES

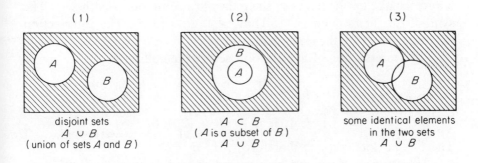

(1)	(2)	(3)
disjoint sets $A \cup B$ (union of sets A and B)	$A \subset B$ (A is a subset of B) $A \cup B$	some identical elements in the two sets $A \cup B$

REVIEW QUESTIONS

1.13. Show what will be the result of the union of the following pairs of sets.
   (a) $P = \{$TOM, JOE, JIM, PETE$\}$
      $Q = \{$HARRY, JIM, JACK, JOE$\}$
   (b) $D = \{20, 40, 50, 60\}$
      $E = \{10, 20, 30, 40\}$
   (c) $J = \{$all numbers $+ 2 = 7\}$
      $K = \{$all numbers $< 2 = 10\}$

1.14. Make up three pairs of sets that will conform to the Venn diagrams in the examples above.

1.15. Show the result of the union of the following sets.
(a) $A = \{1, 3, 5\}$ $B = \{1, 3\}$ $C = \{7, 9\}$
(b) $D = \{a, b, c\}$ $E = \{f, g, h\}$ $F = \{b, c, g, h, j, k\}$

1.16. Draw Venn diagrams showing the relationships involved in problem 1.15.

Intersection of Sets

Intersection of two sets are those elements that are common to both sets and those elements are themselves a set.

EXAMPLE

$$A = \{1, 3, 5, 7\}   B = \{3, 5, 9, 11\}$$

3 and 5 make up the new set, since they are common to both sets. The symbol for intersection is $\cap$. Thus, $A \cap B = \{3, 5\}$. The intersection of sets A and B are the elements 3 and 5.

The intersection of two disjoint sets is the null set.

EXAMPLE

$$A = \{1, 3, 5\}   B = \{2, 4, 6\}$$

There is no intersection; therefore; $A \cap B = \varnothing$.

Intersection of sets can be represented by Venn diagrams by shading in the areas of intersection only.

EXAMPLE

All electricians: $A = \{$Jones, Brown, Smith, Black, Green, Johnson$\}$

All employees over $B = \{$Olson, Jones, Brown, Black, James, 50 years. of age: Finn$\}$

Employees in District $C = \{$Jones, Smith, Black, Field, Solo$\}$
1, 3, 5, with salaries
over $5,000:

Before trying to draw any Venn diagrams, let us develop a practical problem that relates to the example shown here.

Assume that a company has a personnel file (on punched cards) containing the following information.

DISTRICT	EMPLOYEE NO.	NAME	AGE	BASE PAY RATE	GROSS	FICA	FED TAX

NET	REGULAR HRS.	OVERTIME	TOT. HRS. THIS PERIOD	JOB CODE

The company has a thousand employees, so there are a thousand cards in the file, one for each employee. The universal set would then be a thousand (all employees of the company). Cards in the file can be sorted in many ways to furnish information to management.

EXAMPLES

Sort out all cards with job code electrician.
Sort out all cards with age 50 or over.
Sort out all employees in Districts 1, 3, and 5 whose gross pay is over $5,000.

 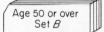  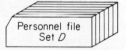

You can see the possible interrelationships of these different sets. Some electricians (set *A*) will also be in set *B* and set *C*; some aged 50 or over (set *B*) will also be in sets *A* and *C* and some in set *C* will be in sets *A* and *B*. New sets can be formed of the cards that are common to the other sets. This can become very confusing without some simple way of showing the relationships.

Set *E* = all in set *A* who are also in set *B*.
Set *F* = all in set *A* who are also in set *C*.
Set *G* = all in set *A* who are also in both sets *B* and *C*.

Now we have a total of six subsets of set *D* and all subsets have some interrelationships with each other.
Returning to the example on page 10, we see that:

Jones and Brown are common to sets *A* and *B*.
Jones and Black are common to sets *B* and *C*.
Jones, Smith, and Black are common to sets *A* and *C*.
Jones is common to all three sets.

New sets are formed from these elements:

$$E = \{\text{Jones, Brown}\}$$
$$F = \{\text{Jones, Smith, Black}\}$$
$$G = \{\text{Jones}\}$$

Sets *A* and *B*

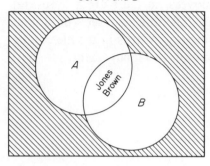

Sets *B* and *C*

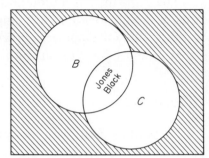

Sets *A* and *C*

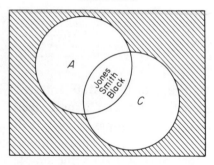

Sets *E* and *F*

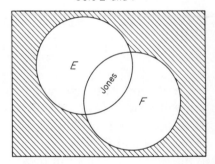

Sets *E* and *G*

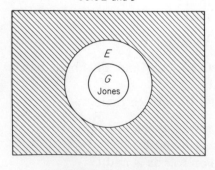

Sets *F* and *G*

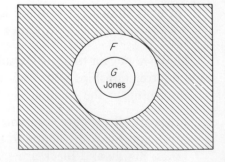

A simple way of showing intersection of sets is with a matrix, or partial matrix, if the individual elements are unique.

EXAMPLE

	Column A	Column B	Column C	Column D	Column E
Row F	1	2	3	4	5
Row G	6	7	8	9	10
Row H	11	12	13	14	15
Row I	16	17	18	19	20
Row J	21	22	23	24	25

Now consider each row as a set: $F = \{1, 2, 3, 4, 5\}$. Each column also is a set: $A = \{1, 6, 11, 16, 21\}$. The intersection of set F and set A is the element 1. This may be written in one of two ways:

$$F \cap A = \{1\}$$
$$\{1, 6, 11, 16, 21\} \cap \{1, 2, 3, 4, 5\} = \{1\}$$

Both of these statements say, "The intersection of set F and set A is the element 1."

EXAMPLES

In the matrix above:

1. What is the intersection of sets H and D?

$$H \cap D = \{14\}$$

2. What is the intersection of sets G and E?

$$G \cap E = \{10\}$$

3. What is the intersection of sets B and C?

$$B \cap C = \varnothing$$

(There is no intersection; therefore, the answer must be the null set.)

REVIEW QUESTIONS

1.17. What is the intersection of the following pairs of sets?
   (a) $P = \{a, c, e, g\}$ $Q = \{g, m, c, x\}$
   (b) $A = \{a, b, c, d\}$ $B = \{1, 2, 3\}$
   (c) $F = \{$names of second six months of year$\}$
      $G = \{$names of third quarter months of the year$\}$

1.18. Draw Venn diagrams showing the intersection of sets in problem 1.17.

1.19. How many sets are there in problems 1.17(a), (b), and (c)?

1.20. Given sets $A = \{1, 3, 5, 7\}$, $B = \{3, 7, 9, 11\}$, $C = \{1, 12, 13, 15\}$. What is the intersection of the following?
(a) $A \cap B$
(b) $B \cap C$
(c) $A \cap C$

1.21. Draw a Venn diagram showing the existing relationships for A, B, and C in problem 1.20.

1.22. How many sets are there in problem 1.20?

Complement Sets

The *complement* of a set may be defined as all elements of a universal set that are not contained in the original set.

EXAMPLE

$$A = \{\text{Monday, Wednesday}\}$$

The total universe of this set are the seven days of the week. The complement of A would be the remaining five days not included in set A. Complement is written with a bar over the symbol:

$$\bar{A}$$

The example, then, would be: $A = \{\text{Monday, Wednesday}\}$, $\bar{A} = \{\text{Tuesday, Thursday, Friday, Saturday, Sunday}\}$. Another way to write the same thing is:

$$\overline{\{\text{Monday, Wednesday}\}} = \{\text{Tuesday, Thursday, Friday, Saturday, Sunday}\}$$

Showing the complement of a set with a Venn diagram:

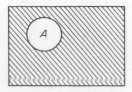

The entire rectangle represents the universe; therefore, the whole shaded area represents $\bar{A}$.

EXAMPLE

$$P = \{1, 2, 3, 4\}$$
$$\bar{P} = \{5, 6, 7, \ldots\}$$

The complement of P would be all counting numbers other than the ones contained in P. In this case, the universe is infinite (shown by the "...").

REVIEW

Review of Symbols

$\{x, y\}$ (braces)	are placed on either side of elements of a set
...	indicates an infinite set
$\{\ \}$ or $\varnothing$	null or empty set
$\in$	is an element of a set
$\notin$	is not an element of a set
$\subset$	is a subset of a set
$=$	set X contains the following elements; two sets are identical
$\Longleftrightarrow$	indicates equivalence; i.e, two or more sets are equivalent
$\cup$	union of two sets
$\cap$	intersection of two sets
$\bar{A}$ (bar over symbol)	complement of a set

Review of Terminology

1. Sets may be:

finite	have definite limits
infinite	endless
null	have no elements
universal	the total environment within which a set (or sets) exists
disjoint	have no elements in common
identical	two (or more) sets contain exactly the same elements
equivalent	one-to-one relationship between sets

2. A subset contains some of the elements of a larger set.

3. A set may contain some of the elements of another set and other nonrelated elements.

4. The union of two sets is obtained when every element of two sets is contained in a third set.

5. The intersection of two sets are all elements that are common to both sets.

6. The complement of a set contains all elements of the universal set not included in the original set.

REVIEW QUESTIONS

1.23. Given the following three sets: $A = \{1, 3, 4, 6\}$, $B = \{2, 5, 7\}$, $C = \{1, 3, 6, 7, 8\}$. Show the elements for the following.
(a) $A \cup B = ?$
(b) $A \cap C = ?$
(c) $B \cap C = ?$
(d) $(A \cup B) \cap C = ?$

1.24. Show the relationships in the following sets, using symbols wherever possible.
(a) $K = \{a, b, d\}$ $L = \{a, d\}$
(b) $M = \{1, 2, 3\}$ $N = \{1, 2, 3\}$
(c) $O = \{6, 5, 4\}$ $P = \{1, 2, 3\}$
(d) P is an element of set G

1.25. Show the complement of the following sets.
(a) $R = \{$Monday, Wednesday, Friday$\}$
(b) $P = \{1, 3, 5, 7\}$
(c) $\overline{A} = \{$alphabet a through $m\}$
(d) $B = \{$six marbles out of 20$\}$

1.26. Draw Venn diagrams to reflect the following sets.
(a) $H = \{a, b, c, d, e\}$ $I = \{b, d, f, h\}$
(b) $J = \{a, b, c, d\}$ $K = \{b, c, d\}$
(c) $L = \{g, h, j, k, p, s\}$ $M = \{j, k, p, s\}$ $N = \{p, s\}$
(d) $O = \{a, c, e\}$ $P = \{b, d, f\}$
(e) $Q = \{a, c, g, h\}$ $R = \{c, d, e, f, g\}$ $S = \{a, b, c, d, j, k\}$

1.27. Show the intersection of sets:
(a) O and P in problem 1.26(d)
(b) Q, R, and S in problem 1.26(e)

2 NUMBER SYSTEMS

BACKGROUND

Some sort of counting activity has been carried out by man since his earliest developmental days. In the very early days, sticks and stones were used as aids and counting devices. The fingers were also very handy and were extensively used by early civilizations. The Pueblo Indians used the fingers of one hand only, creating a *base-five* number system, while the Maya Indians used both hands and both feet, creating a *base-twenty* system.

The Romans appreciated the ten-finger counting system so much that their entire method of counting was based on it (Fig. 2.1).

Many of our present concepts about numbers are based on Roman usage. For example, the word *digit* is derived from the Latin *digitus*, meaning *finger*. Our present day decimal system is based on ten digits much as the early Romans based their number system on the ten fingers of the two hands.

The greatest difference between our system and that of the Romans is that the Romans did not recognize zero as a digit. This forced them to develop a different symbol for every different number they wished to express in writing. (For example, 10 could have been expressed as IO instead of X.)

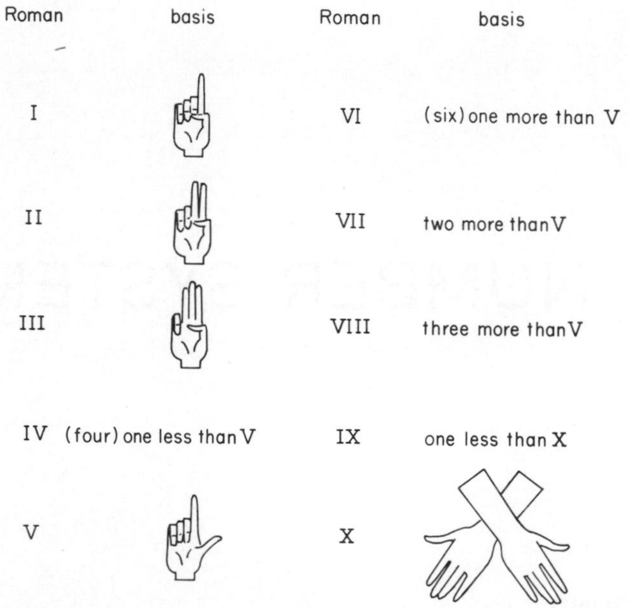

Roman	basis	Roman	basis
I		VI	(six) one more than V
II		VII	two more than V
III		VIII	three more than V
IV (four) one less than V		IX	one less than X
V		X	

Fig. 2.1. Roman numbering system.

The study of computers and data processing requires some knowledge of number systems, since it is number systems that provide the basis for all computer operations. Writing numbers, counting and performing the four basic operations of arithmetic (addition, subtraction, multiplication, division) have been developed by man. In nearly all countries of the western world, a single number system is used. This system is called the *Hindu-Arabic system of numeration*. Although this system has been in use for many centuries, the use of zero was not known until the ninth century A.D.

It was the addition of the zero as a digit in the Hindu-Arabic system that provided the versatility needed to make the system really practical, eliminating the awkwardness and repetition of systems such as the Roman, Egyptian, and other ancient systems.

DECIMAL NUMBER SYSTEM

The most important factor in the development of science and mathematics has been the invention of the decimal number system. It has ten symbols: 0, 1, 2, 3, 4, 5, 6, 7, 8, 9, which are usually called *Arabic numerals*. Counting in tens probably resulted from the fact that man has ten fingers.

The *base* (also called *radix*) of any number system is the number of different digits it possesses including the zero. The decimal system is a *base-ten* system. The use of 10 as the base of a system is not important in itself. Any standard base would do just as well. For example, in our own society the *duodecimal* (base-12) system is commonly used in clocks, inches, dozens, and other counting devices and systems.

Although the concept of zero greatly simplified counting and the manipulation of numbers, another concept is equally important; the concept of *position*. This means that the *value* of each digit is determined by its *position*.

EXAMPLE

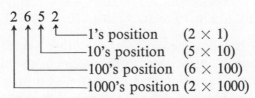

The 2 in the 1000's position has a different value than the 2 in the 1's (units) position. We express this value difference when we say the number: two thousand six hundred fifty two.

Thus, each digit in a sequence of numbers has a *digit value* and a *place value*.

digit value 2, place value 1000
↓   ╭─ digit value 6, place value 100
2 6 5 2 ◄── digit value 2, place value 1
   ↑
digit value 5, place value 10

Notice also that the positioning of the digits increases by a power of 10 as you move to the left. The units position has a value of 1; the tens position has a value of 10×1; the hundreds position has a value of $10 \times 10 \times 1$; etc. The number in the previous example can now be written:

$$2652 = 2 \times (10 \times 10 \times 10 \times 1) + 6 \times (10 \times 10 \times 1)$$
$$+ 5(10 \times 1) + 2 \times 1$$

Although the above example is correct, it is not very practical. A shorthand form is commonly used, called *exponentiation*. The exponent simply indicates how many times the *base* should be included as a factor in the expression being written.

$$2652 = 2 \times 10^3 + 6 \times 10^2 + 5 \times 10^1 + 2 \times 10^0$$

$$
\begin{aligned}
2 \times 10^0 &= &2 \\
5 \times 10^1 &= &50 \\
6 \times 10^2 &= 6 \times (10 \times 10) &600 \\
2 \times 10^3 &= 2 \times (10 \times 10 \times 10) = &2000 \\
\hline
& &2652
\end{aligned}
$$

To summarize:

10^0 simply means 1
10^3 means $10 \times 10 \times 10 \times 1$
10^5 means $10 \times 10 \times 10 \times 10 \times 10 \times 1$

The *superscript* is called the *power* of the other number.

Every person is introduced to the decimal number system very early in his formal education and he continues through simple arithmetic as a matter of course, but the concepts that apply to the decimal system also apply to other number systems which are necessary in working with digital computers.

The following definitions can be derived from the preceding discussion.

1. A number system is a means of indicating the number of *units* counted.

2. The *base* (or radix) of a number system is the number of symbols it has, including the zero.

3. All modern number systems include the *zero*.

4. A *number* is an arbitrary symbol that represents some given quantity of units.

5. A *unit* is the standard by which counting is accomplished.

6. The term *quantity* refers both to a *unit* and a *number* (of units).

REVIEW QUESTIONS

2.1. What was the basis for the Roman number system?

2.2. In what way did this differ from our present-day number system?

2.3. What is the first digit in the decimal number system?

2.4. What is the *base* of the decimal number system?

2.5. What name is given to the decimal number system?

2.6. When was zero first introduced into the decimal system?

2.7. Explain the concept of position in a counting system.

2.8. What is the meaning of the term "exponent"?

2.9. Write the following decimal numbers using powers of 10 to express each number.

    (a) 47   (c) 2685

    (b) 123   (d) 14,203

Note: The correct exponent can quickly be found by simply counting the number of positions to the right of the digit under consideration.

BINARY NUMBER SYSTEM

Some time has been devoted discussing counting on the fingers and the decimal number system, but this has a definite bearing on the method of counting for computers. Since the fingers are a basic counting device, every principle that applies to the fingers also applies to more sophisticated counting devices.

To return to the ten fingers, it can be mathematically proven that the most efficient code for a base-ten system is one that doubles the value of each finger (Fig. 2.2).

This code is often called the *8-4-2-1 code* after the four lowest numbers in the sequence. This code can be used to count from zero to a maximum of 1,023; including zero, this makes a total of 1,024 numbers.

$$512 + 256 + 128 + 64 + 32 + 16 + 8 + 4 + 2 + 1$$
$$= 1023 + 1 = 1024$$

Modern digital computers also use the 8-4-2-1 code. Nearly all computers work in the *binary* mode, which is a base-two system utiliz-

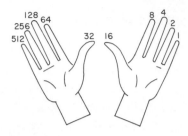

Fig. 2.2. The 8-4-2-1 code.

ing only two digits, *zero* and *one*. This is most convenient for computers because a transistor is either conducting or cut off; a switch is either open or closed; current is flowing in one direction or the other; voltage is either positive or negative; and so on. These are also base-two types of actions. Since computers use binary circuits, the internal arithmetic of computers is binary in nature.

The importance of number systems other than decimal is not immediately apparent to most people. We are so accustomed to using the decimal system that it has become almost second nature to us, while other number systems seem strange and difficult. You will find that they are difficult only because they are strange.

In reality, the three systems to be discussed in this chapter (binary, octal, and hexadecimal) are quite important to computer programmers, and, after a bit of study, you will realize that each system is as simple as the decimal system. The new ground rules must be understood before the systems fall into place in a logical manner.

In the binary system, only two digits are used, zero and one. It requires the invention of a code (using only zero and one) that will cover all possible combinations of numbers to have a workable system. An arbitrary code could easily be devised, but we want a very efficient code and for this reason the 8-4-2-1 code will be utilized (Fig. 2.2). Details of this code and some of the methods for using it will be covered in the following pages.

Counting in the Binary System

The 8-4-2-1 code described on the previous pages is, in fact, the binary code. It utilizes just two digits, zero and one. In such a *two-state* code, the *one* is arbitrarily chosen as the *on* condition and the *zero* as the *off* condition. Therefore, only the *ones* will be counted in a sequence of binary numbers.

The location of each of the ones (based on the 8-4-2-1 code) is the key to its value, as shown in Fig. 2.3. It is important to note that

Code value →	512	256	128	64	32	16	8	4	2	1	Count only ones
	0	0	0	0	0	1	0	1	0	0	= 20 (16 + 4)
	0	0	0	0	1	0	0	1	0	1	= 37 (32 + 4 + 1)
	0	0	0	0	0	1	1	0	1	1	= 27 (16 + 8 + 1)

Fig. 2.3. Code values of the 8-4-2-1 code.

the values given to each binary position start from the *right* and progress to the *left*. (If this sequence were reversed, the results would be entirely different.) The rightmost position, then, is the position of the *least-significant digit* (LSD) and the leftmost position is that of the *most-significant digit* (MSD). Notice that this is exactly the same as in the decimal number system, in which the leftmost position is the most significant (Fig. 2.4).

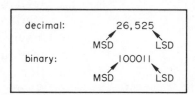

Fig. 2.4. Position significance of numbers.

A few examples should assist in the conversion of decimal numbers to binary and binary to decimal.

EXAMPLES

What is the decimal value of each of the following binary numbers?

1.   0 1 0 1 1 0   *Step 1:* Find the *posi-*
           ↓   ↓ ↓   *tional* value of each of the
Values: 16   4 2   1's.

$$16 + 4 + 2 = 22$$

2.
```
   1  0  0  1  0  1     1
   ↓        ↓     ↓     ↓
  64   +    8  +  2  +  1 = 75
```

(See code value in Fig. 2.3.)

Step 2: Add the *values* together.

Note that the positional value of binary numbers double as you move to the left. Each value to the left is twice the previous value to the right.

3.
```
   1  0  0  1   0   1  0  1  0  1  0
   ↓        ↓       ↓     ↓     ↓
 1024  +  128  +  32  +  8  +  2 = 1194
```

Convert the following decimal numbers to binary.

1. 61_{10}

Step 1: Find the positional value just equal to or lower than the number you want to convert. Mark a "1" for this position.

Values: $\boxed{32}$
       1

Step 2: Look at the next lower position. If it is added to the value you already have, will it be equal to or less than the total you want? If so, mark a 1. If it adds up to a number that is too large, mark a zero for that position.

Values: $\boxed{32}\boxed{16}$
       1 1 (totals 48)

Step 3: Continue this process until you reach the desired total. Place zeros in all unused positions.

Values: $\boxed{32}\boxed{16}\boxed{8}\boxed{4}\boxed{2}\boxed{1}$
       1 1 1 1 0 1
       $32 + 16 + 8 + 4 + 1 = 61$

2. 28_{10} *Values:* $\boxed{16}\boxed{8}\boxed{4}\boxed{2}\boxed{1}$
              1 1 1 0 0

3. 9_{10} *Values:* $\boxed{8}\boxed{4}\boxed{2}\boxed{1}$
             1 0 0 1

A number of other codes based on the binary system are possible, and many such codes are used for computers. The 8-4-2-1 code described above is usually called the *pure binary code*.

REVIEW QUESTIONS

2.10. What is the most efficient code for a ten-position counting system?

2.11. Computers work in a base-two mode, which is usually called _____.

2.12. Why is the base-two mode so convenient for computers?

2.13. What digits are utilized in the base-two numbering system?

2.14. In the binary code, which digit represents the "on" condition and which digit represents the "off" condition?

2.15. What is the value of each of the following sequences of binary numbers?
(a) 001101   (c) 001010000
(b) 0001111100   (d) 001000111

2.16. Show the following numbers in binary notation.
(a) 48   (c) 13
(b) 27   (d) 128

2.17. What is the 8-4-2-1 code usually called?

Since binary numbers tend to be extremely long (roughly 3.3 times longer than decimal numbers), it is more convenient to group them in threes. This does not change the relative value of each position but simply makes it easier to read.

Values:

256	128	64		32	16	8		4	2	1

```
       001            010        100
        ↙              ↓           ↘
        64    +        16    +     4   =    84
```

Counting in the binary system is as follows:

Decimal	Binary
0	000
1	001
2	010
3	011
4	100
5	101
6	110
7	111
8	001 000
9	001 001

Since the binary system contains only 0 and 1, it is necessary to take the same "move" at 2 that is taken at 10 in the decimal system. This is to place a "1" to the left and start again (at the right) with "0." Therefore, a decimal 2 is a binary 10, 3 is 11, and then another shift must be made, adding 1 to the left and starting again with 0.

EXAMPLES

Convert the following decimal numbers to binary.

1. $22 = 010 \quad 110$

$16 + 4 + 2 = 22$

2. $76 = 001 \ 001 \ 100$

$64 + 8 + 4 = 76$

In the 8-4-2-1 sequence, find the number just less than the value of the one to be converted. Start with this number, and continue adding until the required number is reached. Add zeros to the left of the MSD to complete this last (leftmost) group of three binary digits. The zeros are added just to keep all groups in threes.

REVIEW QUESTIONS

2.18. Show the binary equivalents of the following decimal numbers.

(a) 6

(b) 9

(c) 15

(d) 32

(e) 73

(f) 426

2.19. Convert the following binary numbers to decimal.

(a) 011 101

(b) 010 001 101

(c) 111 111

(d) 000 011 001

By this time, you should have noticed the similarities between the decimal and binary number systems. The decimal system is base-ten. The binary system is base-two. Positional moves in binary, then, are

Table 2.1. Binary-to-Decimal Conversion Table

	Binary-to-Decimal Conversion										
Power value	2^{10}	2^9	2^8	2^7	2^6	2^5	2^4	2^3	2^2	2^1	2^0
Decimal value	1024	512	256	128	64	32	16	8	4	2	1

by the power of two, just as they are by the power of ten in decimal. All rules that apply to decimal also apply to binary. Study the conversion table in Table 2.1.

Binary Arithmetic, Addition

Only a few rules need to be observed to accomplish simple arithmetic in binary form:

ADDITION:

> Rule 1: Zero plus zero equals zero.
> Rule 2: Zero plus one equals one. (One plus zero also equals one.)
> Rule 3: One plus one equals zero with a *carry* of one to the left.

Adding two binary numbers is very much like adding two decimal numbers. See the examples below:
Add 276 and 345.

$$
\begin{array}{rl}
11 & \text{(carry)} \\
276 & \text{(addend)} \\
\underline{345} & \text{(augend)} \\
621 & \text{(sum)}
\end{array}
$$

Whenever the sum of one pair of digits is greater than 9, only the units digit is recorded in the "sum" column and a carry of one moves to the next position to the left. The addition process in binary is much the same, as shown by the following examples:

EXAMPLES

1. Add 15 + 7.

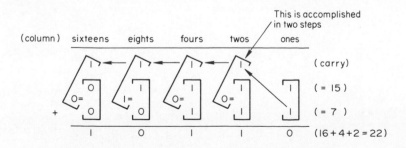

In the ones column, Rule 3 applies. In the twos column, Rule 3 applies again, but we must further add the carry, so the result is 1 with a carry. The same thing happens in the fours column. In the eights column, Rule 2 applies, but again we must add the carry; so now Rule 3 takes over, and we end up with zero and a carry. In the sixteens column, Rule 1 applies, then add the carry, which winds it up with a 1.

2. Add 3 + 3.

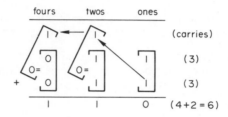

In the ones column, Rule 3 applies. In the twos column, two steps must be taken: first, $1 + 1 = 0$ with a carry; second, the 0 (resulting from the first step) $+ 1$ (from the previous carry) $= 1$. In the fours column, two steps must be taken: first, $0 + 0 = 0$; second, the $0 + 1$ (from the previous carry) $= 1$. Each time there is a carry, the second step must be taken.

3. Add 4 + 3.

fours	twos	ones	
			(carries)
1	0	0	(4)
+ 0	1	1	(3)
1	1	1	
			(4+2+1 = 7)

4. Add $7 + 6$.

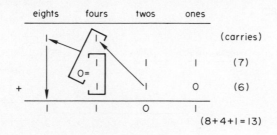

eights	fours	twos	ones	
				(carries)
		I	I	(7)
+		I	O	(6)
I	I	O	I	(8+4+1=13)

REVIEW QUESTIONS

2.20. State the three rules for binary addition.

2.21. Work the following problems.

(a)

sixteens	eights	fours	twos	ones	
O	I	O	O	I	(9)
+ O	O	I	I	O	(6)

(b)

sixteens	eights	fours	twos	ones	
I	I	O	O	I	(25)
+ O	O	I	I	O	(6)

(c)

sixteens	eights	fours	twos	ones	
	I	O	I	O	(10)
+	O	I	I	I	(7)

Binary Arithmetic, Subtraction

The rules for binary subtraction are as follows:

SUBTRACTION:

Rule 1: Zero minus zero equals zero.
Rule 2: One minus one equals zero.
Rule 3: One minus zero equals one.
Rule 4: Zero minus one equals one, with one borrowed from the left.

EXAMPLES

1. Subtract 15 − 7.

(column)	sixteens	eights	fours	twos	ones	
						(borrows)
	O	I	I	I	I	(=15)
−	O	O	I	I	I	(= 7)
	O	I	O	O	O	(= 8)

Applying the rules above, in the ones column, Rule 2 applies, and also, in the twos and fours columns. In the eights column, Rule 3 applies. In the sixteens column, Rule 1 applies.

2. Subtract 12 − 4.

	eights	fours	twos	ones	
					(borrows)
	I	I	O	O	(12)
−	O	I	O	O	(4)
	I	O	O	O	(= 8)

In these examples, only Rules 1, 2, and 3 were used. Now we will try an example that uses Rule 4.

3. Subtract 12 − 7.

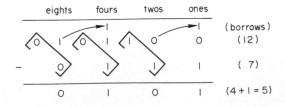

	eights	fours	twos	ones	
		I	I		(borrows)
	O	I	O	O	(12)
−	O	I	I	I	(7)
	O	I	O	I	(4 + 1 = 5)

In the ones column, Rule 4 applies; but, since there is no 1 to borrow in the twos column, we must get it from the fours column, changing the 1 to a 0 in the fours and the 0 to a 1 in the twos. In the twos column, Rule 2 applies. In the fours column, 0 − 1 causes a borrow from the eights column, leaving it a 0, which results in 0 for the final subtraction.

Similar but somewhat different rules are used for multiplication and division. They are nothing more than sequences of addition and subtraction, extremely cumbersome with paper and pencil, but very rapidly accomplished with the high speeds attained by modern computers. These examples demonstrate the way arithmetic is actually accomplished within the computer. Binary arithmetic is interesting because it shows the method used by the computer, but it is not necessary to memorize the rules since you would have no occasion to perform arithmetic in such a slow and cumbersome manner.

Complement Arithmetic

Since multiplication can be accomplished by series of additions and division can be accomplished by series of subtractions, it follows that computers only need circuitry to accomplish the two basic functions of addition and subtraction.

However, this is not true, because the technique of *complement* arithmetic allows the computer to have just *one* type of circuitry and still accomplish all of the basic arithmetic operations.

Subtraction can be accomplished by the addition of complements. Let us explore the meaning of this sentence first in the decimal system. The complement of 10 is the difference between 10 and any given number.

EXAMPLES

The 10's complement of 4 $10 - 4 \ \ = 6$
The 10's complement of 7 $10 - 7 \ \ = 3$
The 10's complement of 35 $100 - 35 \ = 65$
The 10's complement of 362 $1000 - 362 = 638$

Another commonly used system is the 9's complement, which is the difference between 9 and any given number.

EXAMPLES

The 9's complement of 4 $9 - 4 \ = 5$
The 9's complement of 7 $9 - 7 \ = 2$
The 9's complement of 48 $99 - 48 = 51$

Now we will try a few subtraction problems using both 10's and 9's complements.

Subtract 42 from 68.

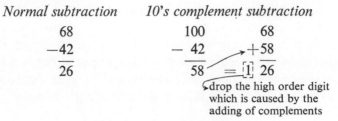

Normal subtraction	10's complement subtraction

```
Normal subtraction          10's complement subtraction
      68                       100          68
    −42                      −  42        +58
    ────                     ────         ────
      26                       58   = [1] 26
```

↳drop the high order digit
which is caused by the
adding of complements

9's complement subtraction

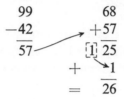

```
       99             68
      −42           +57
      ────          ────
       57          [1] 25
                 +     1
                 ────────
                 =     26
```

In 9's complement, the extra digit is not dropped, but added to the units position. This is called *end-around carry*.

Subtract 3 from 7.

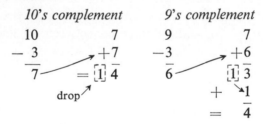

```
   10's complement        9's complement
     10        7          9           7
   − 3      +7          −3          +6
   ───      ────        ──          ────
     7  = [1] 4          6          [1] 3
        drop                      +     1
                                  ─────────
                                  =      4
```

REVIEW QUESTIONS

2.22. Show the 10's complement subtraction of the following decimal numbers.

(a)   7   (b)   246

    −2   −127

2.23. Show the 9's complement subtraction of the following decimal numbers.

(a)   8   (b)   4562

    1   −1379

In the binary system, 1's complement and 2's complement are used. These are quite simple because there are only two numbers to work with: 0 and 1. To complement a binary number, using 1's complement, simply change all zeros to ones and all ones to zeros.

EXAMPLE

0 0 1 1 0 1 0 1 1	Binary number
1 1 0 0 1 0 1 0 0	Complement

In the computer, this can be accomplished in the following manner: Subtract 1001 from 1111.

$$(9) \qquad (15)$$

$$
\begin{array}{r}
1\ 1\ 1\ 1 \\
-\ 0\ 1\ 1\ 0 \quad \text{(complement of 1001)} \\
\hline
1\ 0\ 0\ 1 \\
1\ 1 \quad \text{(carry)} \\
\hline
[1]\ 0\ 1\ 0\ 1 \\
\longrightarrow 1 \quad \text{(end-around carry)} \\
\hline
0\ 1\ 1\ 0 \quad \text{(decimal 6)}
\end{array}
$$

This may seem difficult to you, but it is very simple for the computer. The 2's complement can be obtained by adding one to the units position of the complement.

OCTAL NUMBER SYSTEM

We have said that the binary system is a *base-two* system. The octal number system is a *base-eight* system and is very convenient to use as a shorthand to binary.

Since it is a base-eight system, it will utilize only the numerals 0 through 7. Counting in this system is as shown in Fig. 2.5 (notice that 8 and 9 are never used):

When writing a number in octal, it is usual to designate the system in the following manner: 376_8. The *subscript* 8 indicates that 376 is an octal number.

DECIMAL	OCTAL	DECIMAL	OCTAL	DECIMAL	OCTAL
0	0	8	10	16	20
1	1	9	11	17	21
2	2	10	12	18	22
3	3	11	13	19	23
4	4	12	14	20	24
5	5	13	15	21	25
6	6	14	16	22	26
7	7	15	17	23	27

Fig. 2.5. Decimal-to-octal conversion.

As in the decimal and binary systems, the positional value of each digit in a sequence of numbers is definitely fixed.

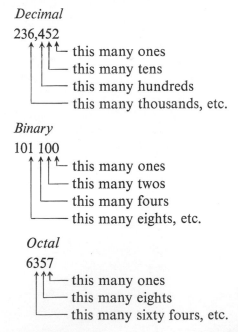

Decimal

236,452
— this many ones
— this many tens
— this many hundreds
— this many thousands, etc.

Binary

101 100
— this many ones
— this many twos
— this many fours
— this many eights, etc.

Octal

6357
— this many ones
— this many eights
— this many sixty fours, etc.

The base-eight (octal) system can be demonstrated using the powers of numbers, as shown below:

$$375_8 = 3 \times 8^2 + 7 \times 8^1 + 5 \times 8^0$$
$$1246_8 = 1 \times 8^3 + 2 \times 8^2 + 4 \times 8^1 + 6 \times 8^0$$

Before discussing the details of the octal system, it may be interesting to examine some of the different number systems. These are shown in Table 2.2 and you will notice that you now understand two systems: the base-ten (decimal) and the base-two (pure binary).

Table 2.2. **Names of Number Systems with Bases from 2 Through 20**

Base	Name
2*	pure binary
3	ternary
4	quaternary
5	quinary
6	senary
7	septenary
8*	octal (octonary)
9	novenary
10*	decimal
11	unidecimal
12	duodecimal
13	terdenary
14	quaterdenary
15	quidenary
16*	hexadecimal (sexadecimal)
17	septendecimal
18	octodenary
19	novendenary
20	vicenary

*Systems covered in this chapter.

REVIEW QUESTIONS

2.24. What is the base of the octal number system?

2.25. Convert the following decimal numbers to octal.

(a) 6   (d) 17
(b) 8   (e) 20
(c) 12

2.26. What are the *values* of the three low-order positions in the following number systems?

(a) decimal
(b) binary
(c) octal

Conversion between Octal and Binary

The relationship between octal and binary is so simple that conversion may be made instantaneously. Consider every binary number in groups of threes (001010101 = 001 010 101). Now, each grouping of three binary digits is idetified by ones, twos, and fours positions, and these are used to convert to octal.

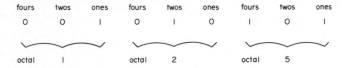

Conversely, each octal number is converted to three binary numbers as in Fig. 2.6. The binary representation of the octal numbers is often called *binary-coded octal*.

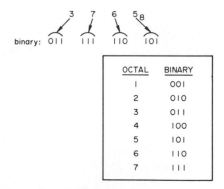

OCTAL	BINARY
1	001
2	010
3	011
4	100
5	101
6	110
7	111

Fig. 2.6. Octal-to-binary conversion.

REVIEW QUESTIONS

2.27. Convert the following binary numbers to octal.

(a) 101

(b) 010 110

(c) 110 101 111

(d) 001 011 100

2.28. Convert the following octal numbers to binary.

(a) 27_8 (c) 1267_8

(b) 450_8 (d) $34,165_8$

Converting from Octal to Decimal

This is usually accomplished by looking up the number in an Octal-D conversion table (refer to Appendix B). It may be accomplished manually in the following manner:

Multiply each octal position in turn by eight, starting with the leftmost position. Then add the next number to the result, multiplying the sum by eight, and continue until the last digit is reached. This one is not to be multiplied.

EXAMPLES

1.

$$3327_8 = ?_{10}$$

$$
\begin{array}{r}
3327_8 \\
\times\ 8 \\
\hline
24 \\
+\ 3 \\
\hline
27 \\
\times\ 8 \\
\hline
216 \\
+\ 2 \\
\hline
218 \\
\times\ 8 \\
\hline
1744 \\
+\ 7 \\
\hline
1751_{10}
\end{array}
$$

Result: $(3327_8 = 1751_{10})$

2.

$$426_8 = ?_{10}$$

$$
\begin{array}{r}
426_8 \\
\times\ 8 \\
\hline
32 \\
+\ 2 \\
\hline
34 \\
\times\ 8 \\
\hline
272 \\
+\ 6 \\
\hline
278_{10}
\end{array}
$$

Result: $(426_8 = 278_{10})$

REVIEW QUESTIONS

2.29. Convert the following octal numbers to decimal notation. Do the first three problems manually, then look up the others in Appendix B.

(a) 17_8 (e) $1{,}407_8$

(b) 56_8 (f) $7{,}214_8$

(c) 133_8 (g) $10{,}000_8$

(d) 560_8 (h) $10{,}123_8$

Note: Use of the Octal-Decimal Conversion Table is explained on page 39.

Converting from Decimal to Octal

This procedure is also generally accomplished by checking a conversion table, but it may be done manually in the following manner:

Successively divide the decimal figure by eight, until no further division is possible. The *octal result* will be the last quotient figure, followed by each of the remainders, starting from the last and finishing with the first.

EXAMPLES

1.

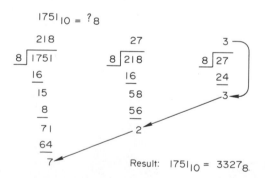

2.

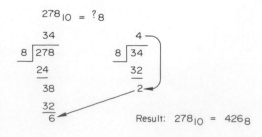

3.

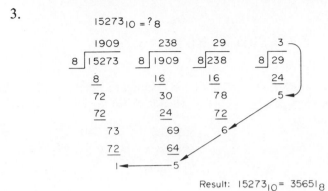

$$15273_{10} = ?_8$$

Result: $15273_{10} = 35651_8$

Use of the Octal-Decimal Conversion Table

You will note, when examining this table, that the octal numbers are along the top and down the left-hand side of the table. Reading the table can most readily be explained with an example.

EXAMPLE

Assume that the number to be converted is 154_8. Procedure: Find 150 in the left column and 4 in the top column. The intersecting point is the decimal equivalent number (in this case, 108).

The numbers in the first column (under the octal 0) are the equivalent decimal numbers to the octal numbers in the left-hand column.

To convert in the opposite direction (from decimal to octal), simply find the desired number in the body of the table, then add the octal numbers in the left-hand column and top row for the desired answer.

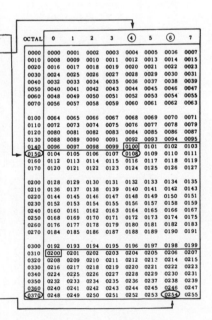

EXAMPLE

Convert decimal 254 to octal. Find 254 in the table. The number in the left-hand column is 370 and the number in the top row is 6, totaling 376 for the octal answer.

REVIEW QUESTIONS

2.30. Convert the following numbers from decimal to octal notation. Work the first three problems manually; then look up the rest of the answers in Appendix B.

(a) 8_{10} (e) 999_{10}

(b) 29_{10} (f) 3003_{10}

(c) 357_{10} (g) 4096_{10}

(d) 579_{10}

2.31. Convert the following octal numbers to decimal. Use the conversion table, Appendix B.

(a) 70_8 (d) 276_8

(b) 117_8 (e) 1526_8

(c) 344_8

HEXADECIMAL NUMBER SYSTEM

We have discussed decimal, binary, and octal number systems in some detail. These general discussions were developed to lead into computer codes. In some systems, three, four, or six binary digits make up a number or character that the computer can recognize. In the hexadecimal system, four binary characters make up a recognizable character. The *hexadecimal system* must be discussed in some detail because the IBM System/360 computers use this number system. Internally, it uses the binary code. The hexadecimal (HEX.) system is a *base-16* system. The decimal digits 0 through 9 are used as the first ten digits (just as in the decimal system), followed by the letters A through F, which represent the values of 10, 11, 12, 13, 14, and 15. Examine Table 2.3 for a comparison of the three number systems:

A *bit* (*binary digit*) is the smallest unit of information used in the computer. It is represented by a zero (off) or a one (on). You will notice immediately the resemblance to the binary 0 and 1. In fact, they are the same—binary digits are called bits.

A *byte* consists of eight bits in consecutive sequence. Bytes are successive and do not overlap each other in computer memory. On the IBM System/360 computers, a byte is eight bits. We will not discuss the six-bit byte at this time.

Table 2.3. Hexadecimal Conversion Table

Decimal	Binary	Hexadecimal
0	0 0 0 0	0
1	0 0 0 1	1
2	0 0 1 0	2
3	0 0 1 1	3
4	0 1 0 0	4
5	0 1 0 1	5
6	0 1 1 0	6
7	0 1 1 1	7
8	1 0 0 0	8
9	1 0 0 1	9
10	1 0 1 0	A
11	1 0 1 1	B
12	1 1 0 0	C
13	1 1 0 1	D
14	1 1 1 0	E
15	1 1 1 1	F

EXAMPLES

$$\boxed{0\,|\,1\,|\,1\,|\,0\,|\,1\,|\,0\,|\,1\,|\,0}$$ one byte = 8 bits

$$\boxed{0\,|\,0\,|\,1\,|\,0\,|\,1\,|\,0\,|\,0\,|\,0\,|\,1\,|\,1\,|\,0\,|\,0\,|\,0\,|\,1\,|\,0\,|\,1}$$ two bytes = 16 bits

A hexadecimal digit may be thought of as half of a byte. Two hexadecimal digits make up one byte, the rightmost four bits constituting one hexadecimal digit and the leftmost four bits constituting another hexadecimal digit.

EXAMPLES

One byte:

$$\boxed{0\,|\,0\,|\,1\,|\,0\,|\,1\,|\,0\,|\,1\,|\,1}$$

$\underbrace{}_{2}\quad\underbrace{}_{11(\text{decimal})\,=\,B}$

$$\boxed{1\,|\,1\,|\,1\,|\,1\,|\,0\,|\,0\,|\,1\,|\,0\,|\,0}$$

$\underbrace{}_{E}\quad\underbrace{}_{4}$

As you can see, there is a little translation involved here. First, you must translate the bits to their decimal equivalents, then (if it is above 9) a second translation to the hexadecimal equivalent. In time, you can learn to translate directly into hexadecimal. *Note*: With so many different number systems, you must learn to be specific about which system you are using. The system may be indicated by using a *subscript* of the *base* you are using.

EXAMPLES

1. Binary: $1\ 0\ 1\ 1_2$ ←— base 2

2. Decimal: $2\ 0\ 8_{10}$ ←— base 10

3. Hexadecimal: $7\ E_{16}$

4. Hexadecimal: $4\ 6_{16}$ ←— base 16

Usually, no subscript is used for binary because it is so obviously binary (a long string of ones and zeros), but it is wise to indicate the number system when using other systems.

REVIEW QUESTIONS

2.32. How many bits to a byte on a System/360 computer?

2.33. How many hexadecimal digits to a byte?

2.34. Convert the following hexadecimal digits to binary.
  (a) E   (c) B
  (b) 9   (d) F

2.35. Convert the following binary numbers to hexadecimal digits.
  (a) 1011   (c) 1010
  (b) 0100   (d) 1101

EBCDIC (*E*xtended *B*inary *C*oded *D*ecimal *I*nterchange *C*ode)

The EBCDIC code was developed for several reasons. The BCD code has a total capacity of 64 characters (six-bit code) while EBCDIC (eight-bit code) can accommodate 256 characters, making it much more flexible and capable of handling many more characters. Also, the basic

Hollerith keypunch code was retained in EBCDIC, virtually without change.

A total of 256 characters can be represented by this code. It is accomplished with eight bits ($2^8 = 256$), which is equivalent to two HEX. digits. The 256 possible characters (each of which can be represented by a single byte) are known as the EBCDIC code.

The 16×16 matrix shown below (Fig. 2.7) contains the entire code.

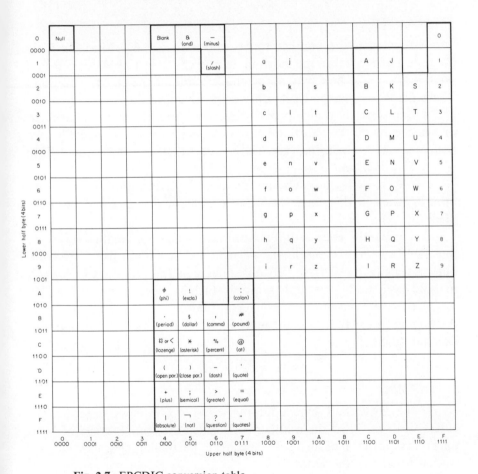

Fig. 2.7. EBCDIC conversion table.

Notice that both the HEX. numbers and the equivalent binary (four bit) representation are shown as the labels for the rows and columns.

The columns (vertical) represent the high-order four bits, or HEX. digit, and the rows (horizontal) represent the low-order four bits, or HEX. digit.

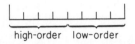

high-order   low-order

EXAMPLE

Assume that the number to be converted is $C6_{16}$. Find the C in the upper half byte at the bottom of the chart in Fig. 2.7; then find the 6 in the lower half byte along the left side of the chart. The point where the two converge is the answer; in this case it is F.

To convert in the opposite direction, find the digit. character, or symbol in the body of the table; then locate the applicable two half bytes in the margins. The HEX. equivalent for the symbol dash (–) is: 6D. This can also be shown in bits: 0 1 1 0 1 1 0 1 .

So far, only the more commonly used characters and symbols have been assigned in this code. Assigned characters are enclosed in heavy black borders.

The EBCDIC code is quite important as it is used extensively with the IBM System/360 computers. A printout (or listing) from the computer is usually in HEX., so that it is quite regularly necessary to convert to more easily understandable symbols, such as those shown in Fig. 2.7.

A complete hexadecimal conversion table is included as Appendix A, ranging from HEX. 000 through HEX. FFF (which is equivalent to decimal 4095).

EXAMPLE

If the lower half byte contains 0 0 1 1 , it could refer to the letters C, L, or T, or the number 3. It takes the combination of both upper and lower half bytes to specify a particular character. If we wished to specify the letter "L":

$$1\ 1\ 0\ 1\ 0\ 0\ 1\ 1$$

upper   lower
D   3

REVIEW QUESTIONS

2.36. What character is represented by the following HEX. digits?

(a) 40   (d) E2

(b) 7B   (e) 5C

(c) C7   (f) F8

2.37. What would be the byte representation of the following characters?

(a) A   (c) , (comma)

(b) &   (d) F

2.38. Convert the following HEX. numbers to decimal. (Use the conversion table, Appendix A.)

(a) 1C4   (d) AEB

(b) 40A   (e) E7F

(c) 88D

OTHER CODING SYSTEMS

There is a large variety of computer codes in use today. In fact, two different computers seldom use exactly the same code in their internal operation. On the other hand, it is not practical for anyone to try to memorize dozens of different codes.

There are three basic categories of codes:

1. *Regularly weighted codes.* In these codes, each number has a weighted value which is regular and is based on a specific rule. If you know the rule, you can determine the weight of any number in the sequence. The 8-4-2-1 code is an example of this type of code. It increases progressively by a power of 2.

2. *Arbitrarily weighted codes.* In these codes, each number is weighted, but without a regular rule. The weights are arbitrarily assigned. For these codes, you must know the assigned values.

3. *Nonweighted codes.* In these codes, the numbers have no weighted values. Each one in the series is defined to represent a specific quantity. The Roman number system is an example of nonweighted codes.

Table 2.4 shows examples of commonly used codes:

Table 2.4. Commonly Used Codes

Decimal	Binary*	Binary-Coded Decimal	(BCD)	7421 Code	Biquinary Code	Excess-3 Code (XS-3)
0	000	000	000	0000	01 00001	0011
1	001	000	001	0001	01 00010	0100
2	010	000	010	0010	01 00100	0101
3	011	000	011	0011	01 01000	0110
4	100	000	100	0100	01 10000	0111
5	101	000	101	0101	10 00001	1000
6	110	000	110	0110	10 00010	1001
7	111	000	111	1000	10 00100	1010
	(r.w.)†	(r.w.)†		(a.w.)‡	(r.w.)†	(n.w.)**

*This is identical with binary-coded octal for the first eight digits.
†r.w.=regularly weighted code.
‡a.w.=arbitrarily weighted code.
**n.w.=nonweighted code.

The BCD (or 8–4–2–1) code is very useful and has been much used in computers. The BCD code uses a six-bit byte, whereas the EBCDIC code uses an eight-bit byte.

The XS-3 code is formed by adding 3 to each decimal number and then forming the binary number in the normal way (weighted 8-4-2-1). Refer to Table 2.4 for a comparision of decimal digits and the XS-3 code, The XS-3 code is not a weighted code because the sum of the bits does *not* equal the number that is represented. The XS-3 code has been widely used on many militarized computers.

REVIEW QUESTIONS

2.39. Name the three categories of computer codes.

2.40. Identify the category of code represented by each of the following counters.
   (a) A five-position counter with values 2, 4, 6, 8, 10
   (b) A five-position counter with values 1, 3, 7, 9, 12
   (c) A five-position counter with values as shown:

$$0 = 000$$
$$1 = 001$$

$$2 = 011$$
$$3 = 010$$
$$4 = 110$$

Error Detecting Codes

A computer does not usually give any indication of circuitry problems if they occur. The only sign of circuitry problems is if one notices that expected results do not materialize (in the expected manner).

There are two techniques related to circuitry that are possible to solve the potential error problem: (1) the problem can be run through the computer twice to see if the results are the same both times, and (2) dual circuitry can be built so that the problem can be run through both circuits simultaneously.

Both of these techniques are expensive and time consuming; therefore, a third solution is necessary. The solution taken by most computer manufacturers is to use *error detecting* codes, which contain more information than is absolutely necessary to specify the required digit. The biquinary and 7-4-2-1 codes are in this category.

The 8-4-2-1 code can also be used as an error detecting code by adding one extra bit (called a *parity bit*) to each position of the code. Table 2.5 shows how a parity bit can be added to the 8-4-2-1 code so that each digit will always come out with an *even* number of 1's.

Obviously, a similar parity code can be set up that causes all the digits to come out with *odd* numbers. The odd parity check has the

Table 2.5. Even Parity Check

Decimal	8421	Parity Code	Number of 1's
0	0000	0	0
1	0001	1	2
2	0010	1	2
3	0011	0	2
4	0100	1	2
5	0101	0	2
6	0110	0	2
7	0111	1	4
8	1000	1	2
9	1001	0	2
10	1010	0	2

advantage that zero can never be mistaken for no information (0000 in Table 2.4 above). For this reason, the odd parity check (Table 2.6) is more common than the even parity check.

Table 2.6. Odd Parity Check

Decimal	8421	Parity Code	Number of 1's
0	0000	1	1
1	0001	0	1
2	0010	0	1
3	0011	1	3
4	0100	0	1
5	0101	1	3
6	0110	1	3
7	0111	0	3
8	1000	0	1
9	1001	1	3
10	1010	1	3

This can be carried further by constructing circuitry in such a way that the error (detected by the parity bit) can be corrected as soon as it is detected, thus assuring that calculations will always result in accurate answers. A little additional circuitry is needed for this, but it is well worth the price to be always sure of the result (if the problem has been constructed correctly in the first place).

REVIEW

Review of Symbols

Rules for binary addition:

$$0 + 0 = 0$$
$$0 + 1 = 1$$
$$1 + 0 = 1$$
$$1 + 1 = 0 \text{ with a carry of 1 to the left}$$

Rules for binary subtraction:

$$0 - 0 = 0$$
$$1 \quad 1 - 0$$
$$1 - 0 = 1$$
$$0 - 1 = 1 \text{ with 1 borrowed from the left}$$

Review of Terminology

Base (radix)	of a number system is the number of symbols it has, including zero
Superscript	small number above a symbol or number, indicating the power of that quantity (e.g., A^2, A^1, A^0)
Quantity	refers to both a unit and a number of units
Binary	base-two number system
Octal	base-eight number system
Hexadecimal	base-sixteen number system
Subscript	a small number below a symbol or number, indicating the number system (base) being represented (e.g., 45_8, 150_{10})
Bit	binary digit
Byte	eight bits in consecutive sequence
Parity bit	an extra bit added to a computer code to aid in verifying accuracy

REVIEW QUESTIONS

2.41. What is the base of the following?
   (a) binary number system
   (b) octal number system
   (c) hexadecimal number system

2.42. Show the binary equivalents of the following decimal numbers.
   (a) 4   (d) 25
   (b) 7   (e) 69
   (c) 12

2.43. Convert the following binary numbers to decimal.
   (a) 010 110   (c) 010 000 101
   (b) 001 101 110   (d) 011 001 001

2.44. State the three rules for binary addition.

2.45. Convert the following binary numbers to octal.
   (a) 010 110   (c) 001 011 100
   (b) 010 101 110   (d) 011 010 111

2.46. Convert the following octal numbers to binary.
   (a) 16   (c) 47
   (b) 22   (d) 253

2.47. In the binary code, which digit represents the following?
   (a) *on* condition
   (b) *off* condition

2.48. What is a *byte* (System/360)?

2.49. How many hexadecimal digits are there in one byte?

2.50. Convert the following decimal numbers to their hexadecimal equivalents.
   (a) 4   (d) 14
   (b) 11   (e) 10
   (c) 8

3 REVIEW
OF ALGEBRA

This chapter is included as a brief review of some of the basic rules and principles of algebra. Much of the material may be familiar to the student, but merely being able to work a problem is not enough if the rule that leads to the correct solution has been forgotten. For this reason, some of the basic principles and important rules and concepts will be given, followed by short examples and problems, to encourage the remembering process.

OPERATORS

An *operator* is a mathematical symbol which represents a mathematical process to be performed on one or more associated *operands*.

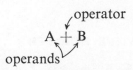

There are three kinds of operators: arithmetic, logical and relational. Each of the three has its own symbology.

1. *Arithmetic* operators.

   Add $+$

   Subtract $-$

   Multiply $\times$ If two letter symbols are to be multiplied, the operator is often omitted (e.g., $A \times B$ is the same as AB).

   Divide $\div$ (also $/$)

   Exponent (the power of a number)   This is written as a superscript (e.g., B^0, B^1, B^2, B^3, etc.).

2. *Logical* operators.

   AND $\cdot$

   OR $+$

   NOT $-$ (or $'$) (e.g., $\overline{A}$ means Not A)

These are used extensively in Boolean algebra and in symbolic logic. They will be discussed in Chapters 4 and 5.

3. *Relational* operators. These refer to expressing a relationship between two items.

   Greater than $>$

   Less than $<$

   Equal to $=$

   Unequal $\neq$

   Less than or equal to $\leqslant$

   Greater than or equal to $\geqslant$

EXAMPLES

   $A > B$ A is greater than B

   $A < B$ A is less than B

   $A = B$ A is equal to B

   $A \neq B$ A is not equal to B

   $A \geqslant B$ A is greater than or equal to B

POSITIVE AND NEGATIVE NUMBERS

The plus and minus signs are used to indicate positive and negative numbers.

EXAMPLES

+27 or 27   (If not negative, the + sign may be omitted.)

−27   (Negative numbers should always be preceded by the minus sign.)

Algebraic Rules for Addition

1. Adding two plus numbers results in the sum of the two numbers with a sign of plus.

2. Adding two minus numbers results in the sum of the two numbers with a sign of minus.

3. Adding two numbers with different signs (+, −) results in the *difference* between the two numbers and carries the sign of the larger number.

$$
\begin{array}{ccccc}
+2 & -2 & +2 & +3 & +2 \\
+2 & -2 & -2 & -1 & -6 \\
\hline
+4 & -4 & 0 & +2 & -4
\end{array}
$$

Algebraic Rules for Subtraction

The rules for subtraction are just like the rules for addition except that the sign of the subtrahend is *reversed* and then *added* to the minuend.

Minuend:		$+2$	$=$	$+2$		$+2$	$=$	$+2$		-4	$=$	-4		$+7$	$=$	$+7$
Subtrahend:		$-(+2)$		-2		$-(-2)$		$+2$		$-(+6)$		-6		$-(+2)$		-2
Difference:				0				$+4$				-10				$+5$

Algebraic Rules for Multiplication

1. Multiplying number with like signs results in a sign of plus.
2. Multiplying numbers with unlike signs results in a sign of minus.

$$
\begin{array}{cccc}
+\,2 & -\,9 & -\,9 & +\,9 \\
+\,9 & -\,2 & +\,2 & -\,2 \\
\hline
+18 & +18 & -18 & -18
\end{array}
$$

Algebraic Rules for Division

The rules for division are identical to the rules for multiplication:
1. Division of numbers with like signs results in a sign of plus.
2. Division of numbers with unlike signs results in a sign of minus.

$$
-3\overline{)-6} = +2 \qquad +3\overline{)+6} = +2 \qquad -3\overline{)+6} = -2 \qquad +3\overline{)-6} = -2
$$

REVIEW QUESTIONS

3.1. Name the three types of operators.

3.2. Write the applicable symbols for the following operators.
  (a) Multiply
  (b) Exponent (X to the third power)
  (c) AND
  (d) Greater than or equal to
  (e) Subtract
  (f) Unequal
  (g) OR
  (h) Less than

3.3. Show the results of the following algebraic arithmetic problems.
  (a) Add:

$$
\begin{array}{cccc}
+2 & & & \\
-3 & +3 & -7 & 4 \\
+4 & +4 & -3 & 3 \\
-6 & -2 & -4 & 6 \\
\hline
\end{array}
$$

  (b) Subtract:

$$
\begin{array}{cccc}
+6 & +8 & -6 & -9 \\
-(+3) & -(-3) & -(+4) & -(-4) \\
\hline
\end{array}
$$

  (c) Multiply:

$$
\begin{array}{cccc}
+9 & -8 & +7 & -9 \\
(\times)+6 & (\times)-3 & (\times)-4 & (\times)+5 \\
\hline
\end{array}
$$

  (d) Divide:

$$
+2\overline{)+8} \qquad -3\overline{)-9} \qquad -2\overline{)+10} \qquad +4\overline{)-8}
$$

Successive Arithmetic Operations

When numbers are in successive sequence (several numbers to be operated upon) and more than one operator is involved, special rules prevail.

1. Successive numbers involving only multiplication or addition may be solved in any order without causing a change in the answer.

EXAMPLES

(a) $3 + 6 + 2 + 8 = 19$

Changing the order: $6 + 3 + 8 + 2 = 19$

(b) $2 \times 4 \times 6 \times 3 = 144$

Changing the order: $6 \times 2 \times 4 \times 3 = 144$

2. When multiplication and division are involved with numbers in sequence along with addition and subtraction, the multiplication and division must be performed first.

EXAMPLES

(a) $2 + 6 \times 3 = 20$

If this had been worked sequentially, the answer would have been $2 + 6 = 8 \times 3 = 24$, which is wrong (when working problems with algebra).

(b) $2 + 4 \div 2 = 4$

Wrong way: $2 + 4 = 6 \div 2 = 3$

(c) $3 + \underbrace{6 \times 3} + 2 - \underbrace{1 \times 2} =$

$3 + \quad 18 \quad + 2 - \quad 2 \quad = 21$

Wrong way: $3 + 6 = 9 \times 3 = 27 + 2 = 29 - 1 = 28 \times 2$
$= 56$

3. If parentheses are used in an expression or equation, they should be handled first.

EXAMPLES

(a) $2 + \underbrace{(3 \times 2)}6 =$

$2 + \quad \underbrace{6 \times 6} =$

$2 + \quad\quad 36 \quad = 38$

(b) $96 - 4\underbrace{(3 + 2)}(4) =$

$96 - \underbrace{4 \times 5} \times 4 =$

$96 - \quad \underbrace{20 \times 4} =$

$96 - \quad\quad 80 = 16$

4. A number multiplied by zero gives a zero result. A number cannot be divided by zero.

EXAMPLES

(a) $6 \times 0 = 0$
(b) $4 \times 2 \times 0 = 0$
(c) $8 \times 3 \times 0 = 0$

REVIEW QUESTIONS

3.4. What is the sequence used in solving an equation (as it relates to the arithmetic operators)?

3.5. Solve the following problems.
(a) $5 + 3 \times 6 + 4 =$
(b) $5 \times 2 + 6 \div 3 + 2 =$
(c) $5(2 + 3) + 6 \times 2 =$
(d) $(2 \times 3) + (6 \times 4) =$
(e) $5 + 2(6 \div 3) \times 0 =$
(f) $3(8 \div 2) \times 0 =$

FRACTIONS

A *fraction* is a way of saying that a quantity has been divided into a number of different parts. The *numerator* indicates the number of parts and the *denominator* indicates the number of divisions in the total quantity.

$\dfrac{3}{4}$ three parts of four $\dfrac{4}{4}$ four parts of four (which equals 1)

1. Both the numerator and denominator of a fraction may be multiplied or divided by the same number without changing the value of the fraction.

EXAMPLES

(a) $\frac{2}{3} \times \frac{2}{2} = \frac{4}{6} \times \frac{3}{3} = \frac{12}{18} \times \frac{4}{4} = \frac{48}{72}$ (reduced $= \frac{2}{3}$)
(b) $\frac{24}{30} \div \frac{2}{2} = \frac{12}{15} \div \frac{3}{3} = \frac{4}{5}$

All of these fractions are exactly the same. The rule for multiplication and division does not carry over to addition and subtraction.

WRONG EXAMPLE

$$\tfrac{3}{4} + \tfrac{1}{1} = \tfrac{4}{5}$$

This changes the value entirely.

2. If fractions are to be added or subtracted, they must first be reduced to a common denominator. Then the numerators only are added or subtracted.

EXAMPLES

(a) $\tfrac{3}{4} + \tfrac{1}{3} + \tfrac{1}{2} = \tfrac{9}{12} + \tfrac{4}{12} + \tfrac{6}{12} = \tfrac{19}{12}$

(b) $\tfrac{3}{4} - \tfrac{1}{6} - \tfrac{1}{3} = \tfrac{9}{12} - \tfrac{2}{12} - \tfrac{4}{12} = \tfrac{3}{12}$ (reduced $= \tfrac{1}{4}$)

3. Multiplication of fractions is quite simple. Merely multiply the numerators and then multiply the denominators separately. This automatically reduces the fractions to a common denominator.

EXAMPLE

$$\frac{1}{3} \times \frac{2}{5} \times \frac{5}{6} = \frac{1 \times 2 \times 5}{3 \times 5 \times 6} = \frac{10}{90} = \frac{1}{9}$$

4. To divide fractions, invert the divisor and multiply according to Rule 3 above.

EXAMPLE

$$\frac{3}{4} \div \frac{1}{3} = \frac{3}{4} \times \frac{3}{1} = \frac{3 \times 3}{4 \times 1} = \frac{9}{4}$$

Decimal Fractions

Decimal fractions always have a denominator of 10, or a power of 10 (100, 1000, 10,000, etc.). Common fractions can be written as decimals by dividing the numerator by the denominator.

EXAMPLES

(a) $\frac{4}{10} = .4$, $\frac{4}{100} = .04$, $\frac{3}{1000} = .003$

(b) $\frac{3}{5} = 5\overline{)3.0} = .6$

(c) $\frac{2}{3} = 3\overline{)2.00} = .66\ldots$ (The three dots mean that the number can be carried to infinity.)

1. When adding or subtracting decimal numbers, the decimal points must be kept in line, one with the other.

EXAMPLES

26.3	.4
2.45	.321
.342	0.5
12.6	2.01
41.692	3.231

2. When multiplying numbers that have decimal places, the answer will have as many decimal places as the sum of the decimal places in the multiplier and multiplicand.

EXAMPLES

$$\begin{array}{r} 2.45 \\ \times\ \ .06 \\ \hline .1470 \end{array}$$ (two decimal places)
(two decimal places)
(four decimal places)

$$\begin{array}{r} 3.052 \\ 2.32 \\ \hline 6104 \\ 9156 \\ 6104 \\ \hline 7.08064 \end{array}$$

$$\begin{array}{r} 2.04 \\ .04 \\ \hline .0816 \end{array}$$

3. When dividing with decimal fractions, the decimal point is placed in the quotient, as many places to the right of the decimal point in the dividend as there are decimal places in the divisor.

EXAMPLES

$$.3\overline{)2.14} \qquad .45\overline{)27.56} \qquad .02\overline{)3.275}$$

REVIEW QUESTIONS

3.6. Multiply the following fractions by 3.
   (a) $\frac{2}{3}$
   (b) $\frac{4}{5}$

3.7. Divide the following fractions by 2.
   (a) $\frac{12}{14}$
   (b) $\frac{22}{30}$

3.8. Work the following problems.
   (a) $\frac{1}{3} + \frac{4}{5} + \frac{1}{4} =$ (c) $\frac{2}{3} \times \frac{1}{4} \times \frac{3}{5} =$
   (b) $\frac{2}{3} - \frac{1}{6} - \frac{1}{4} =$ (d) $\frac{4}{5} \div \frac{1}{6} =$

3.9. Show the following fractions as decimal fractions.
   (a) $\frac{5}{10}$ (b) $\frac{3}{4}$ (c) $\frac{1}{6}$ (d) $\frac{4}{1000}$

PROPORTIONS AND PERCENTAGES

Proportions and percentages are very closely related to each other. The rules will, therefore, be developed for both operations.

1. A given number may be found to be a proportion of a sum or total number, by dividing the number by the total.

EXAMPLES

(a) What proportion of 60 is the number 6?

$$60 \overline{)6.0} = .1 \text{ (proportion)}$$

(b) What proportion of 40 is the number 16?

$$40 \overline{)16.0} = .4$$

2. To translate a proportion to a percent, multiply the proportion by 100.

EXAMPLES

(a) Example (a) above:

$$.1 \times 100 = 10\%$$

(b) Example (b) above:

$$.4 \times 100 = 40\%$$

(c) What percent of 80 is the number 20?

$$80 \overline{\smash{\big)}\,20.0} = .25 \text{ (proportion)}$$
$$= 25\%$$

Note that to get the percentage figure, you simply move the decimal point two places to the right.

(d) What percent is the proportion .6?

$$.60. = 60\%$$

3. To find the number that a given proportion of a total equals, multiply the total by the proportion.

EXAMPLE

Only .3 of 20 people attended a meeting. How many attended?

$$.3 \times 20 = 6 \text{ people}$$

4. The sum of all percents of a given total must always equal 100. The sum of all proportions always equals 1.0.

REVIEW QUESTIONS

3.10. (a) The number 7 is what proportion of 35?
(b) What percentage is the number 7?

3.11. Out of 20 people in the class, only six received a grade of A. What proportion of the class received a grade of A? What percentage is that?

3.12. In problem 3.11, what percentage got a grade less than A? What proportion of the class is that?

POWERS OF NUMBERS

In the decimal number system, numbers progress to the right and left of the decimal point by the power of 10. The powers are indicated by superscripts next to the digit or symbol (e.g., 10^1, X^2, A^4, etc.).

1. The positioning of the digits actually specify the power to be used, based on the *position value* of the digit or symbol.

EXAMPLE

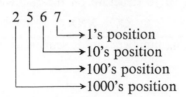

$$2\ 5\ 6\ 7\ .$$

This could also be correctly written:

$$2 \times (1000) + 5 \times (100) + 6 \times (10) + 7 \times (1)$$

The "power" position location simply indicates how many times 10 it must take to reach that particular position.

1's position	0 times
10's position	1 time
100's position	2 times (10×10)
1000's position	3 times $(10 \times 10 \times 10)$

Now it becomes simple to place the superscript:

$$X^6\ X^5\ X^4\ X^3\ X^2\ X^1\ X^0.$$

2. For decimal places, the process is the same, but in the opposite direction.

$$.\ 3\ 2\ 1$$

10*ths* position ◄─┘

100*ths* position ◄───┘

1000*ths* position ◄──────┘

This could also be correctly written:

$$3 \times (\tfrac{1}{10}) + 2 \times (\tfrac{1}{100}) + 1 \times (\tfrac{1}{1000})$$

or
$$3 \times (.1) + 2 \times (.01) + 1 \times (.001)$$
To write the powers:
$$3 \times (10^{-1}) + 2 \times (10^{-2}) + 1 \times (10^{-3})$$

REVIEW QUESTIONS

3.13. Show the power of 10 structure on the following expression.
$$X, XXX. XXX$$
3.14. Write the following numbers out as sequences of powers of 10.
(a) 249
(b) 2.34
(c) 3502

SUMMATION

The word *sum* (meaning to add) is derived from the word summation. It is designated by the symbol Σ, which is the Greek capital letter *sigma*.

EXAMPLE

Assume that X is a *variable*, which may assume a succession of values, such as X for a series of "n" items.
$$\Sigma X = X_1 + X_2 + X_3 + \ldots + X_n$$
The summation of X would require the addition of all the "n" values of X.

1. The summation of a *constant* (a value that does not change) is derived by multiplying the constant by "n," as many times as it (the constant) appears in a series.

EXAMPLE

A is a constant.
$$\Sigma A = nA \ (n \times A)$$

Assume that $A = 4$ and $n = 5$ (A occurs five times).

$$\Sigma A = (5)(4) = 20$$
$$\Sigma A = A_1 + A_2 + A_3 + A_4 + A_5$$

2. The summation of an algebraic sum of two or more items is equal to the algebraic sums of the individual items. This may be shown by formula as follows:

$$\Sigma (A + B + C) = \Sigma A + \Sigma B + \Sigma C$$

EXAMPLE

We will prove the formula with actual values, with $n = 3$ (each item will have three values).

$$
\begin{array}{lll}
A_1 = 2 & A_2 = 3 & A_3 = -4 \\
B_1 = 4 & B_2 = -2 & B_3 = 3 \\
C_1 = 3 & C_2 = -1 & C_3 = 3
\end{array}
$$

$$\Sigma (A + B + C) = \Sigma (2 + 3 - 4) + (4 - 2 + 3) + (3 - 1 + 3)$$
$$= 1 + 5 + 5 = 11$$
$$\Sigma A + \Sigma B + \Sigma C = 1 + 5 + 5 = 11$$

3. The summation of a variable times a constant is equal to the constant times the summation of the variable.

EXAMPLE

Let $C = $ constant (value of 4), $n = 3$. $X = $ variable ($X_1 = 6$, $X_2 = 4$, $X_3 = 3$).

$$\Sigma (CX) \text{ (sum of constant times the variable)}$$
$$\Sigma (CX) = (4 \times 6) + (4 \times 4) + (4 \times 3) = 24 + 16 + 12 = 52$$
$$C \Sigma X \text{ (constant times the sum of the variables)}$$
$$C(\Sigma X) = (4)(6 + 4 + 3) = (4)(13) = 52$$

Therefore:

$$\Sigma (CX) = C \Sigma X$$

4. The summation of a variable divided by a constant is equal to the sum of the variable, divided by the constant. This can be depicted as follows:

$$\Sigma \left(\frac{X}{C} \right) = \frac{\Sigma X}{C}$$

EXAMPLE

$C = \text{constant (value 4)}$, $n = 2$, $X = \text{variable } (X_1 = 8,\ X_2 = 12)$.

$$\Sigma\left(\frac{X}{C}\right) = \frac{8}{4} + \frac{12}{4} = 2 + 3 = 5$$

$$\frac{\Sigma X}{C} = \frac{8 + 12}{4} = \frac{20}{4} = 5$$

REVIEW QUESTIONS

3.15. Define the following terms.
(a) Variable
(b) Constant

3.16. What is the Greek capital letter symbol for "summation"?

3.17. Assume that $A = 6$ and $n = 3$. Show the formula to calculate the summation of A. What is the numeric answer?

3.18. Compute $\Sigma(CX)$ if $C = 3$, $n = 3$, $X_1 = 2$, $X_2 = 3$, $X_3 = 4$.

BASIC EQUATIONS

The most basic rule in dealing with equations is that whatever is done to one side of the equation must also be done to the other side. This rule applies to all four arithmetic operations. If one side is divided by a number or symbol, the other side must also be divided by the same number or symbol.

EXAMPLES

1. *Addition:* $x = X - D$ (add D to both sides)
   Result: $x + D = X$
   Proof: let $x = 2$, $X = 4$, $D = 2$
   $$(x = X - D) \quad 2 = 4 - 2$$
   $$(x + D = X) \quad 2 + 2 = 4$$

2. *Subtraction:* $X = x + D$ (subtract D from both sides)
   Result: $X - D = x$
   Proof: let $x = 2$, $X = 4$, $D = 2$

$$(X = x + D) \quad 4 = 2 + 2$$
$$(X - D = x) \quad 4 - 2 = 2$$

Using numbers to prove equations is fine as a technique to show the relationships, but it is most important to be able to visualize the relationships with symbols only.

3. *Multiplication:*

$$\frac{A}{B} = C \quad \text{(multiply both sides by } B)$$

Result: $A = CB$

4. *Division:*

$$\Sigma X = ND \text{ (divide both sides by } N)$$

Result: $\dfrac{\Sigma X}{N} = D$

5. *Squaring:* (The square of a number is that number multiplied by itself.) $4^2 = 4 \times 4 = 16$, $8^2 = 8 \times 8 = 64$.

$$A = B - C \text{ (square both sides)}$$

Result: $A^2 = (B - C)^2$
A table of squares (to 1000^2) is shown in Appendix C.

These are fundamental principles and a good understanding will make the handling of equations much easier for the student.

SQUARE ROOT

There is no need to go through the fairly laborious job of computing the square root of a number. Square root tables are available to assist in this job. Appendix C is a table of squares and square roots.

To find the square root of any whole number (*integer*) from 1 to 1000, find the number in the column headed N and then read the square root in the column headed $\sqrt{N}$.

EXAMPLES

Find the square root of:
(a) 53   square root = 7.28
(b) 136   square root = 11.662
(c) 613   square root = 24.759

Notice that the square root of 25 is 5. You can see at a glance that 5^2 (5 × 5) equals 25. This is why the table of squares (N^2) is included. This table can be used to find the square root of numbers that are over 1000 or that are not integers.

The number is found in the N^2 column, then the square root is read in the N column.

EXAMPLES

Find the square root of:
(a) 1681   square root = 41
(b) 9216   square root = 96
(c) 100,489   square root = 317

You can prove any of the answers by multiplying it by itself (41 × 41) to reach the original number (1681).

For decimal fractions, it becomes necessary to know where to place the decimal point. Split the number you are working with into two-place groups, going to right and left of the decimal point. If the number is not even, add zeros to the far right or left.

EXAMPLES

1,263.14	12 63 . 14
6.275	06 . 27 50
32,100.1	03 21 00 . 10

There must be one number or decimal place for each pair of digits. Now, follow the same procedure as with the large numbers; find the number (or its nearest approximation) in the N^2 column of the table; then read the square root (or its near approximation) in the N column. Place the decimal point so that the answer will have one digit for each pair in the original number.

EXAMPLES

Find the square root of:
(a) 6.25   (06 . 25)   square root = 2.5
(b) 171.61   (01 71 . 61)   square root = 13.1

(c) 17.12   square root (approximately) $= 4.2$

There is no 1712 in the N^2 column. Take the nearest number, which is 1764, and use it to get the approximate square root.

(d) 908.20   (09 08 . 20)   square root (approximately) $= 30.1$

REVIEW

Review of Symbols

Arithmetic operators:

$+$	add
$-$	subtract
$\times$	multiply
$\div$	divide
A^0, A^1	exponent (superscript)

Logical operators:

$\cdot$	AND⎫ These two are not to be mistaken for the
$+$	OR ⎬ arithmetic operators for multiply and add.
$-$	NOT⎭

Relational operators:

$>$	greater than
$<$	less than
$=$	equal
$\neq$	unequal
$\geqslant$	greater than or equal to
$\leqslant$	less than or equal to
Σ	sigma (summation)

Review of Terminology

Operator	a mathematical symbol representing a mathematical process
Operand	a quantity or symbol to be operated upon by an operator
Square	any number multiplied by itself
Exponent	the power of a number
Sigma	summation

REVIEW QUESTIONS

3.19. What are the names of the logical operators?

3.20. What symbols are used for the relational operators?

3.21. Add the following numbers.

$$
\begin{array}{llll}
\text{(a)} & -8 & \text{(b)} +4 & \text{(c)} -3 & \text{(d)} +3 \\
& +3 & +6 & -9 & -7
\end{array}
$$

3.22. Subtract the following numbers.

$$
\begin{array}{llll}
\text{(a)} & -8 & \text{(b)} +4 & \text{(c)} -3 & \text{(d)} +3 \\
& +3 & +6 & -9 & -7
\end{array}
$$

3.23. Multiply the following numbers.

$$
\begin{array}{llll}
\text{(a)} & -8 & \text{(b)} +4 & \text{(c)} -3 & \text{(d)} +3 \\
& +3 & +6 & -9 & -7
\end{array}
$$

3.24. What is the result of each of the following problems?
(a) $3 + 6 \times 2 - 4 + 3 \times 3 =$
(b) $6 + (4 + 6) - 9 \times 2 =$
(c) $(2 \times 3)(3 \times 4)(4 \times 2) =$

3.25. Multiply the following fractions by 3.
(a) $\frac{1}{3}$ (b) $\frac{3}{5}$ (c) $\frac{2}{7}$

3.26. Add the following fractions.
(a) $\frac{1}{4} + \frac{2}{3} + \frac{1}{2} =$
(b) $\frac{3}{7} + \frac{3}{8} + \frac{1}{4} =$

3.27. Multiply the following fractions.
(a) $\frac{1}{4} \times \frac{2}{3} \times \frac{1}{2} =$
(b) $\frac{2}{3} \times \frac{4}{5} \times \frac{3}{7} =$

3.28. Divide the following fractions.
(a) $\frac{5}{6} \div \frac{2}{3} =$ (b) $\frac{3}{8} \div \frac{1}{7} =$

3.29. How many decimal places will there be if the following sized numbers are multiplied?

$$
\begin{array}{ll}
\text{(a)} \ \text{XXX.XXX} & \text{(b)} \ \text{X.XXXX} \\
\qquad\ \ \text{.XXX} & \qquad\ \ \text{X.XX}
\end{array}
$$

3.30. Convert the following common fractions to decimal fractions (to four decimal places).

(a) $\frac{4}{7}$ (b) $\frac{4}{5}$

3.31. If 12 people, out of a class of 30, received a grade of A, what percentage of the class received A's? What proportion of the class got grades below A?

3.32. Write the following numbers out as sequences of the power of 10.

(a) 165   (b) 3426

3.33. What is the Greek symbol and name for summation?

3.34. Define the terms "constant" and "variable."

3.35. Complete the following formula showing that the algebraic sum of a number of items is equal to the algebraic sums of the individual items.

$$\sum (W + X + Y + Z) = ?$$

3.36. What formula would be the equivalent to the following?

$$\sum (MX) = ?$$

Prove the formula you wrote, using the following.

Constant is: $M = 3, n = 3, X_1 = 3, X_2 = 4, X_3 = 5$

3.37. What is the most basic rule in dealing with equations?

3.38. Find the square roots of the following numbers.

(a) 14   (d) 18.49
(b) 753   (e) 265.225
(c) 3364   (f) 241.63

4 BOOLEAN ALGEBRA

GENERAL BACKGROUND

A knowledge of Boolean algebra is necessary for the understanding of the relationship between the logic of computers and the circuitry incorporated in most computers.

In 1854, George Boole published a book titled *An Investigation of the Laws of Thought On Which are Found the Mathematical Theories of Logic and Probabilities*. His intention was to perform a mathematical analysis of logic. This book was the start of the algebra that bears his name.

Boolean algebra differs considerably from the rules of elementary algebra. Confusion between the two forms may be avoided by considering Boolean algebra from the point of view of switching circuits. In 1938, Claude Shannon first used Boolean algebra on switching circuit problems. He developed a mathematical method of depicting circuits that consisted of switches and relays. His methods were almost universally accepted and are still in use today.

The real advantage of Boolean algebra to the computer designer is that the circuitry used can, with this method, be simplified and expressed in mathematical notation instead of in bulky circuit diagrams. A basic knowledge of Boolean algebra is essential in the computer field.

A number of terms and symbols must be learned by the student who has no background in Boolean algebra. He must also be introduced to basic circuits. Additionally, he must understand that the variables used in Boolean equations have the unique characteristic of being able to assume only one of two possible values: zero and one. No matter how many variables there may be in an equation describing a logical circuit, each variable can only have the value of 0 or the value of 1.

BASIC CIRCUITRY

A drawing of an electrical circuit is called a *schematic diagram*. Symbols are used to represent different parts of the circuit. A few commonly used symbols are shown in Fig. 4.1.

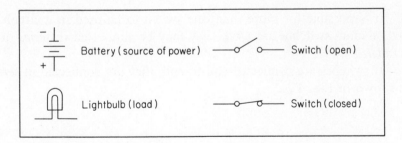

Fig. 4.1. Circuitry symbols.

Most electrical circuits consist of a *source* (source of power), a *switch*, and a *load*. A load is any component that changes electricity into a useful effect. In the example below, the light bulb is the load because the bulb will glow and give off useful light.

EXAMPLE

A complete circuit must have an unbroken path for the current to flow from the negative side of the source through the load and back

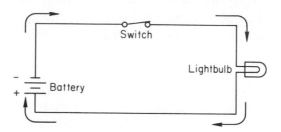

to the positive side of the source. A switch is used to break the circuit.

Current will flow from the source (battery), through the closed switch, through the load (light bulb), and back to the source. If the switch is open, current cannot flow. Therefore, we can say that *output* (the light bulb) will be a 1 if the switch is closed and a 0 if the switch is open.

Series and Parallel Circuits

It is possible for more than one switch to be used in a circuit. If two or more switches are used, they may be connected in *series* or in *parallel*.

If switches are connected end-to-end, they are connected in *series*, as shown in Fig. 4.2.

Fig. 4.2. Switches in series.

It is obvious that both switches must be closed for current to flow through the circuit. If only one switch is closed, current will not be able to flow through the other switch, which is open.

If switches are connected side-by-side, they are connected in *parallel*, as shown in Fig. 4.3.

An examination of this example shows that if either switch is closed, current will flow from the input to the output. Both of these concepts are extremely important to the study of Boolean algebra.

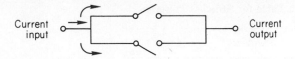

Fig. 4.3. Switches in parallel.

EXAMPLES

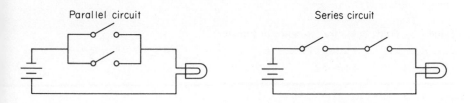

Now we will convert our knowledge of series and parallel circuits to 0 (for no flow of current) and 1 (for flow of current).

How many possible combinations are there for two switches connected in series?

1. Both switches can be open—output 0 (no current).

2. First switch can be open and second one closed—output 0 (no current).

3. Second switch can be open and first one closed—output 0 (no current).

4. Both switches can be closed—output 1 (current flows).

What are the possibilities for two switches connected in parallel?

1. Both switches can be open—output 0.

2. First switch can be open, second one closed—output 1.

3. Second switch can be open, first one closed—output 1.

4. Both switches can be closed—output 1.

For a series circuit, then, the only time that current *can* flow is if both switches are closed. This leads us to the following *truth table* (for a series circuit), Table 4.1, 0 referring to open and 1 referring to closed switches.

In a parallel circuit, the only time that current *cannot* flow is if both switches are open. This leads us to the following truth table (for a parallel circuit), Table 4.2.

Table 4.1. Truth Table for Switches in Series

Switch 1	Switch 2	Output
0	0	0
0	1	0
1	0	0
1	1	1

Table 4.2. Truth Table for Switches in Parallel

Switch 1	Switch 2	Output
0	0	0
0	1	1
1	0	1
1	1	1

REVIEW QUESTIONS

4.1. The development of Boolean algebra is attributed to what man?

4.2. Who first used Boolean algebra to describe switching circuits?

4.3. (a) What values are possible for a single variable in a Boolean equation?
(b) This reminds you of what number system?

4.4. Explain the meaning of *source* and *load*.

4.5. What is the purpose of a switch?

4.6. Draw a schematic of a series circuit.

4.7. Draw a schematic of a parallel circuit.

4.8. What absolute statement can be made about series circuits?

4.9. What absolute statement can be made about parallel circuits?

BOOLEAN SYMBOLOGY

Everyone is familiar with the normal algebraic sign $(+)$ for addition and the dot product sign $(\cdot)$ for multiplication.

$$A + B = C \qquad A \cdot B = C \quad \text{or} \quad AB = C$$
$$5 + 6 = 11 \qquad 5 \times 4 = 20$$

In Boolean algebra, these symbols take on entirely different meanings.

The symbol "$+$" is defined as "OR." Using switches A and B as examples, we can say that either one *or* the other *or* both must be closed for the circuit to conduct current. This is known as the *inclusive OR*. It is shown in Boolean symbology as: $A + B$ (A OR B).

The symbol "$\cdot$" is defined as "AND." This symbol refers to both or all variables connected with the AND sign. If the sign is not there, it is still understood to be AND (e.g., $A \cdot B$ may be written AB and still means A AND B).

Note: It is an unfortunate fact that there is considerable variance in the use of symbology throughout the mathematical community. This is particularly true of the symbols used for the logical AND and OR. The symbols used in this text are felt to be the most commonly accepted by a majority of mathematicians.

To understand these new uses for known symbols, let us return to the circuits for examples.

The AND operation can be represented by switches in *series*:

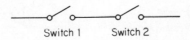

Switch 1   Switch 2

If the switches are open, the output is "0." If the switches are closed, the output is "1." It makes no difference how many switches there are in such a circuit. For current to flow through the circuit, both switch 1 AND switch 2 must be closed. Under any other condition, the output will be "0."

From this we can see that the output of an AND component in a circuit (called a *gate*) is a 1 only if *all* inputs are 1's. In this condition, the gate is considered to be *enabled*. If it is not in this condition (all 1's in; 1 out), the gate is considered to be *disabled*.

The OR operation can be represented by switches in *parallel:*

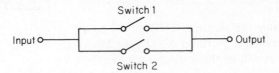

Switch 1

Input o——————o Output

Switch 2

If any switch is closed, the output will be 1. It is called an OR component because when switch 1 OR switch 2 or both switches are closed, the output will be 1. Output will only be 0 if all switches are open.

The number of switches (inputs) has no effect on the result.

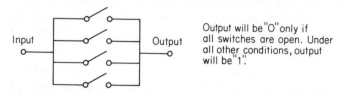

Output will be "0" only if all switches are open. Under all other conditions, output will be "1".

Symbolic representations of AND and OR gates are shown in Fig. 4.4 below:

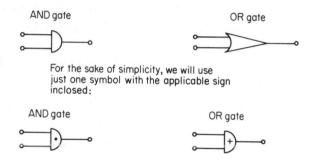

Fig. 4.4. Symbolic representation of AND and OR gates.

REVIEW QUESTIONS

4.10. What is the symbol for Boolean AND?

4.11. What is symbol for Boolean OR?

4.12. Examine the AND gates below. Will the output from each be 0 or 1?

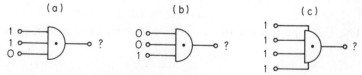

4.13. (a) Enable the pictured AND gate 1 below.
   (b) Disable the pictured AND gate 2 below.
   (c) What will be the output from each?

4.14. Examine the OR gates below. Will the output from each be 0 or 1?

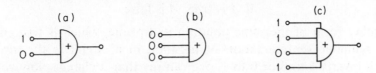

4.15. Enable the pictured symbolic OR gate.
What will be the output?

4.16. What will be the output from each of the following symbolic circuits? Show the output in each phase.

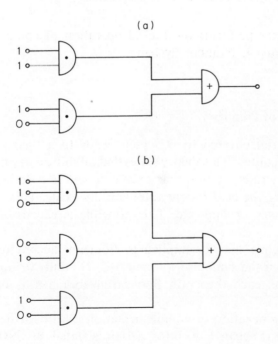

One additional symbol should be mentioned at this time. The negation symbol is a straight line over a character or symbol. In this context, negation signifies the *opposite* of the *existing* state, not a negative number.

EXAMPLE

If A is true, $\bar{A}$ is false.

(Another form of the same notation is A-prime, which is written A'.) The symbolic representation of negation in a circuit is the *inverter*, which inverts 1 to 0 or 0 to 1. We can say that it changes low voltage to high voltage, or high voltage to low voltage. The symbolic representation of an inverter is shown in Fig. 4.5.

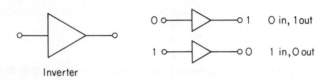

Inverter

Fig. 4.5. Symbolic representation of an inverter.

The inverter performs the logical operation of complementation (refer to Chapter 2, Number Systems).

Principle of Duality

The material covered to this point leads to a theory called the "principle of duality." This theory states that all things may be expressed by two distinct rules, stating that once a system is described by a set of relationships, the *dual system* automatically exists. This dual system is complementary or opposite. This principle is called *De Morgan's theorem*.

This principle may be applied to Boolean algebra to show the balance between the symbols "$+$" and "$\cdot$," "1" and "0," and "A" and "$\bar{A}$." Therefore, each theory in Boolean algebra has its opposite (or complement).

Examining negation or complementation a little further, when an entire quantity is negated, $\overline{A \cdot B}$ (or $\overline{AB}$), it is stated as **NOT A AND B**. If an individual item is negated, $\bar{A} + B$, it is stated as A **NOT** OR **B**.

If A is assumed to be 1, then $\bar{A} = 0$ (the complement). If A is assumed to be 0, then $\bar{A} = 1$. From this we can set up a truth table, as shown in Table 4.3

Table 4.3. Truth Table for NOT
(complement)

NOT Truth Table
$\bar{0} = 1$
$\bar{1} = 0$

Two terms should be introduced at this point: (1) *postulates* are propositions that are taken for granted and (2) *theorems* are rules that can be proven to be true.

EXAMPLES OF POSTULATES

1. $A + 0 = A$ This is taken for granted to be true because whether A is 0 or 1, the postulate is still true.

$$0 + 0 = 0$$
$$1 + 0 = 1$$

2. $A \cdot 1 = A$ As in the example above:

$$0 \cdot 1 = 0$$
$$1 \cdot 1 = 1$$

	(if $A = 0$; $B = 1$)	(if $A = 1$; $B = 0$)
3. $A + B = B + A$	$0 + 1 = 1 + 0$	$1 + 0 = 0 + 1$
4. $B \cdot A = A \cdot B$	$1 \cdot 0 = 0 \cdot 1$	$0 \cdot 1 = 1 \cdot 0$
5. $A + \bar{A} = 1$	$0 + 1 = 1$	$1 + 0 = 1$
6. $\bar{A} \cdot A = 0$	$1 \cdot 0 = 0$	$0 \cdot 1 = 0$

Truth Tables

The truth tables, developed earlier, are repeated in Table 4.4 so that the truth of the postulates and theorems may be more easily checked.

Table 4.4. AND and OR Truth Tables

AND (·) Truth Table

1. 0 AND 0 = 0
2. 0 AND 1 = 0
3. 1 AND 0 = 0
4. 1 AND 1 = 1

OR (+) Truth Table

1. 0 OR 0 = 0
2. 0 OR 1 = 1
3. 1 OR 0 = 1
4. 1 OR 1 = 1

REVIEW QUESTIONS

4.17. What will be the output from each of the following symbolic circuits? Show the output in each phase.

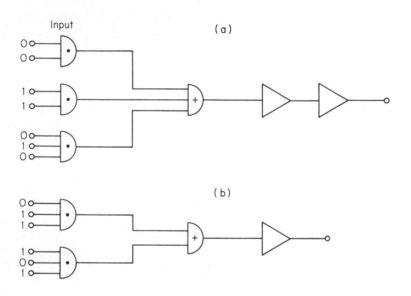

4.18. Translate the following theorems into 1's and 0's and relate them to the proper truth table. Assume that $A = 0$ and $B = 1$.
(a) $A \cdot A = A$
(b) $A(A + B) = A$ (Solve the portion in parentheses first. This is extremely important and it may be of value to check back to Chapter 3, Successive Arithmetic Operations.)

BOOLEAN THEOREMS

A number of basic Boolean algebra theorems are shown in Table 4.5. Some of these have already been discussed.

Table 4.5. Boolean Theorems

1. $A + 0 = A$
2. $A + 1 = 1$
3. $A + A = A$
4. $A + \bar{A} = 1$
5. $A \cdot 0 = 0$
6. $A \cdot 1 = A$
7. $A \cdot A = A$
8. $A \cdot \bar{A} = 0$
9. $(A')' = A$
10. $A + B = B + A$
11. $A \cdot B = B \cdot A$
12. $A + (B + C) = (A + B) + C$
13. $A(BC) = (AB)C$
14. $A(B + C) = AB + AC$
15. $A + AB = A$
16. $A(A + B) = A$
17. $(A + B)(A + C) = A + BC$
18. $A + A'B = A + B$

Explanation of Theorems

Theorem 9 says that double complementation results in the original variable. Variable A can only have the values of 0 or 1. If $A = 0$, then the first complement will be 1 and the second complement will be 0, which is the original value. The same holds true if $A = 1$ originally. Two inverters in a circuit accomplish this purpose, as shown in Fig. 4.6.

Fig. 4.6. Two inverters showing double complementation.

Each theorem can be proven to be true in the same manner. We will examine a few particularly important ones.

Theorems 10 and 11 simply say that the order of addition or multiplication has no effect on the result. Let us first prove these with normal algebra. Later, we will prove them with Boolean algebra, using truth tables to furnish the proof.

EXAMPLES

$$
\begin{array}{cccc}
12 & 18 & 12 & 6 \\
+18 & +12 & \times\ 6 & \times 12 \\
\hline
=30 & =30 & =72 & =72
\end{array}
$$

These two theorems are known as the *commutative laws*.

Theorems 12 and 13 are known as the *associative laws*. Theorem 12 says that when three or more items are to be added, the result will be the same regardless of which two are added together first.

EXAMPLES

Add 12, 16, and 42.

$$
\begin{array}{ccc}
12 & 16 & 12 \\
+16 & +42 & +42 \\
\hline
28 & 58 & 54 \\
+42 & +12 & +16 \\
\hline
=70 & =70 & =70
\end{array}
$$

Theorem 13 is the same as theorem 12, but applied to multiplication. No matter what the sequence of multiplication is, the result will be the same.

EXAMPLES

Multiply 3 × 4 × 6.

$$
\begin{array}{ccc}
3 & 4 & 3 \\
\times 4 & \times 6 & \times 6 \\
\hline
12 & 24 & 18 \\
\times 6 & \times 3 & \times 4 \\
\hline
=72 & =72 & =72
\end{array}
$$

Theorem 14 is called the *distributive law* and it says that if two numbers are to be added, then multiplied by a third number, $[A(B + C)]$, the result will be the same if each of the numbers to be added is multiplied by the third number and then the individual products ($AB + AC$) are added.

EXAMPLE

Formula: $A(B + C) = AB + AC$
Algebraic proof: $5(4 + 3) = (5 \times 4) + (5 \times 3)$
$$5 \times 7 = \boxed{35} \quad 20 \quad + \quad 15 \quad = \boxed{35}$$

Extend the formula and it still works:

$$5(4 + 3 + 2) = (5 \times 4) + (5 \times 3) + (5 \times 2)$$
$$5 \times 9 = \boxed{45} \qquad 20 \quad + \quad 15 \quad + \quad 10 \quad = \boxed{45}$$

All three of the above mentioned laws may be extended to include any number of items. Remember the names: *commutative laws, associative laws,* and *distributive law.* These three laws also apply to normal algebra, as shown in the examples above, but other theorems apply only to Boolean algebra and must be proven to be true in a different manner.

Let us examine theorem 17 as an example:

$$(A + B)(A + C) = A + BC$$

Trying to prove this with normal algebra will be a failure:

$$(5 + 4)(5 + 3) = 5 + (4 \times 3)$$
$$9 \quad \times \quad 8 = \boxed{72} \; 5 + \quad 12 \quad = \boxed{17}$$

To prove this theorem with Boolean algebra, we must first set up truth tables for each side of the formula:

(A + B)(A + C) Table

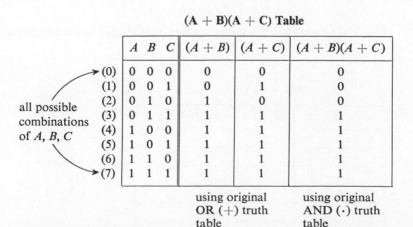

	A	B	C	$(A + B)$	$(A + C)$	$(A + B)(A + C)$
(0)	0	0	0	0	0	0
(1)	0	0	1	0	1	0
(2)	0	1	0	1	0	0
(3)	0	1	1	1	1	1
(4)	1	0	0	1	1	1
(5)	1	0	1	1	1	1
(6)	1	1	0	1	1	1
(7)	1	1	1	1	1	1

all possible combinations of *A, B, C*

using original OR (+) truth table   using original AND (·) truth table

A + BC Table

A	B	C	BC	A + BC
0	0	0	0	0
0	0	1	0	0
0	1	0	0	0
0	1	1	1	1
1	0	0	0	1
1	0	1	0	1
1	1	0	0	1
1	1	1	1	1

Note that the final results in both tables are the same, proving the original theorem.

REVIEW QUESTIONS

4.19. Give a brief definition of each of the following laws.
   (a) Commutative
   (b) Associative
   (c) Distributive

4.20. Set up Boolean truth tables for each side of the following theorem (Theorem 14).

$$A(B + C) = AB + AC$$

The AND and OR, truth tables are reproduced here to simplify the job.

A(B + C) Table

A	B	C	(B + C)	A(B + C)
0	0	0		
0	0	1		
0	1	0		
0	1	1		
1	0	0		
1	0	1		
1	1	0		
1	1	1		

AND (·)
Truth Table

$0 \cdot 0 = 0$
$0 \cdot 1 = 0$
$1 \cdot 0 = 0$
$1 \cdot 1 = 1$

AB + AC Table

A	B	C	A·B	A·C	(AB) + (AC)
0	0	0			
0	0	1			
0	1	0			
0	1	1			
1	0	0			
1	0	1			
1	1	0			
1	1	1			

OR (+)
Truth Table

$0 + 0 = 0$
$0 + 1 = 1$
$1 + 0 = 1$
$1 + 1 = 1$

Relating Theorems to Switching Circuits

How do we relate the Boolean theorems to switching circuits? Let us look at some basic Boolean theorems (rules) and the associated circuitry. Remember that 0 always represents an open switch and 1 represents a closed switch. Switch A may be either open or closed.

1. $A + 0 = A$
2. $A + 1 = 1$

3. $A + A = A$
4. $A + \bar{A} = 1$

(both switches work together)

If A is closed, $\bar{A}$ is open and if A is open, $\bar{A}$ is closed.

5. $A + AB = A$
   Proof: $A + A \cdot B$
   $\quad = A(1 + B)$
   $\quad = A$

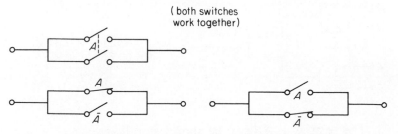

Since the two A switches work together, there is no need for the bottom A or the B switch. The top A switch alone will carry out the same function because even if the B switch were open, current would flow through the upper A switch when A was closed.

Proof of theorem 5 with a truth table:

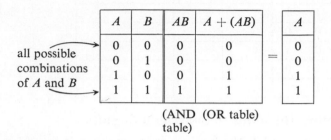

	A	B	AB	$A + (AB)$		A
all possible	0	0	0	0		0
combinations	0	1	0	0	$=$	0
of A and B	1	0	0	1		1
	1	1	1	1		1

(AND (OR table)
table)

Again, we are using the basic AND and OR truth tables.

Examples of the use of symbolic circuits to depict Boolean terms are:

1. The term $A(B + C)$ can be shown as a symbolic circuit as follows.

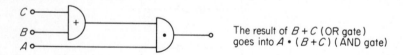

The result of $B + C$ (OR gate) goes into $A \cdot (B+C)$ (AND gate)

2. The term $AB + AC$ can be shown as follows.

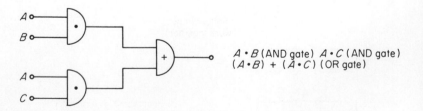

$A \cdot B$ (AND gate) $A \cdot C$ (AND gate)
$(A \cdot B) + (A \cdot C)$ (OR gate)

3. The term $A + AB$ can be shown as follows.

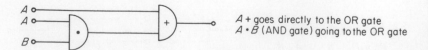

$A +$ goes directly to the OR gate
$A \cdot B$ (AND gate) going to the OR gate

REVIEW QUESTIONS

4.21. Translate the following Boolean terms to symbolic circuits.
  (a) $A + (B + C)$
  (b) $A(BC)$
  (c) $(A + B)(A + C)$

Complement of Boolean Expressions

A simple rule may be followed to find the complement of a Boolean expression: Change all $+$ signs to $\cdot$, all $\cdot$ signs to $+$, and replace each letter in the expression by its complement.

EXAMPLE

$$A(A + B) = A \quad \text{basic expression}$$

complement: $\boxed{\bar{A} + (\bar{A} \cdot \bar{B}) = \bar{A}}$

Now, develop truth tables to prove both theorems:

A(A + B) = A
Truth Table

A	B	$(A + B)$	$A \cdot (A + B)$		A
0	0	0	0		0
0	1	1	0	$=$	0
1	0	1	1		1
1	1	1	1		1
		(using $+$ table)	(using $\cdot$ table)		

Ā + (Ā · B̄) = Ā
Truth Table

$\bar{A}$	$\bar{B}$	$(\bar{A} \cdot \bar{B})$	$\bar{A} + (\bar{A} \cdot \bar{B})$		$\bar{A}$
1	1	1	1		1
1	0	1	1	$=$	1
0	1	1	0		0
0	0	0	0		0
		(using $\cdot$ table)	(using $+$ table)		

SIMPLIFICATION OF BOOLEAN EXPRESSIONS

Boolean expressions can be graphically depicted and simplified with the use of Vietch diagrams. A Vietch diagram is a form of Venn diagram (Chapter 1) that uses blocks to show the intersection of elements.

A Vietch diagram for three variables is shown in Fig. 4.7 below.

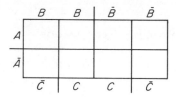

Fig. 4.7. Vietch diagram for three variables.

An X is marked through a square at the intersection of three elements.

EXAMPLES

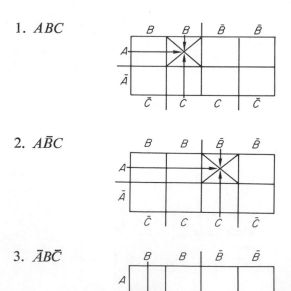

1. ABC

2. $A\bar{B}C$

3. $\bar{A}B\bar{C}$

Figure 4.8 is the basic format for four variables.

The X is drawn in the square that intersects the four variables.

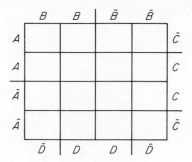

Fig. 4.8. Vietch diagram for four variables.

EXAMPLES

1. $A\bar{B}C\bar{D}$

2. $\bar{A}\bar{B}\bar{C}D$

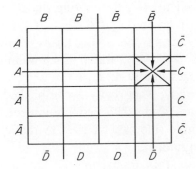

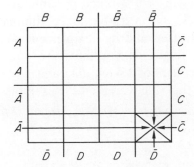

In the same general manner, Vietch diagrams can be drawn for
two variables, five variables, six variables, or as many variables as are
needed for the particular application.

We have shown earlier that computer circuitry (AND gates, OR
gates, and inverters) can be expressed in Boolean equations. Take an

equation such as the following and place each variable into its proper position in a Vietch diagram.

$$K = ABC + A\bar{B}C + \bar{A}\bar{B}C + \bar{A}BC$$

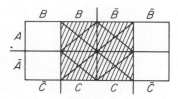

The boxes are compared and boxes with two or more variables in common are reduced to an expression containing only the *common* variables. The first two variables have A and C in common ($ABC + A\bar{B}C$). This is reduced to AC, eliminating the B and $\bar{B}$. The next two terms have $\bar{A}$ and C in common ($\bar{A}\bar{B}C + \bar{A}BC$), reducing to $\bar{A}C$, again eliminating B and $\bar{B}$ from these two terms. Now compare the two simplified terms (AC and $\bar{A}C$), and C is common to both, eliminating A and $\bar{A}$. This reduces finally to $K = C$.

The reduction above was really accomplished with the equation, not the Vietch diagram, and such a reduction is rather tedious and time consuming. One simple rule can be used to reduce equations directly from the Vietch diagram.

Rule: In a Vietch diagram (consisting of any number of variables), two adjacent blocks can be combined to eliminate one variable. When combining two blocks, the variable to be eliminated will be the one which contains both itself and its complement. (There are other rules for more complex situations, but these are not essential to a basic understanding of the uses of Vietch diagrams.)

Let us see how this rule works on the example below.

$$K = ABC + A\bar{B}C + \bar{A}\bar{B}C + \bar{A}BC$$

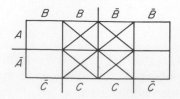

This can be combined in a number of different ways:

The blocks must be adjacent, top and bottom or side by side.

1.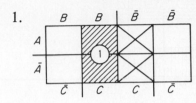

Combining the two darkened blocks will eliminate A and $\bar{A}$ from the two terms involved (ABC and $\bar{A}BC$).

$$K = BC + A\bar{B}C + \bar{A}\bar{B}C$$

2.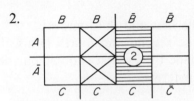

Combining the two darkened blocks will eliminate A and $\bar{A}$ from the other two terms ($A\bar{B}C + \bar{A}\bar{B}C$).

$$K = BC + \bar{B}C$$

3.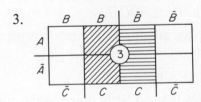

Combining the two new blocks (formed by the other two combinations) will eliminate B and $\bar{B}$ from all of the terms.

$$K = C$$

It is easy to see from this example how quickly simplification can be accomplished with Vietch diagrams.

EXAMPLE

Draw a Vietch diagram and incorporate the following equation.

$$K = A\bar{B}\bar{C}D + A\bar{B}C\bar{D} + AB\bar{C}\bar{D} + ABCD + \bar{A}B\bar{C}D + \bar{A}B\bar{C}\bar{D}$$

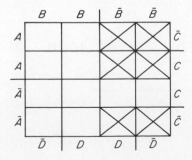

In the previous example, we combined up and down boxes. This time we will combine boxes side-by-side. It really makes no difference as the result will be the same in either case.

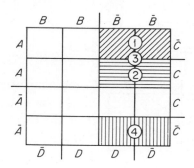

1. Eliminates D and $\bar{D}$ from the first and third terms.
2. Eliminates D and $\bar{D}$ from the second and fourth terms.
3. Combining the new boxes made by the first two combinations eliminates C and $\bar{C}$ from all four terms.

At this point, the formula stands:

$$K = A\bar{B} + \bar{A}\bar{B}\bar{C}D + \bar{A}\bar{B}C\bar{D}$$

4. Combining the final two boxes eliminates D and $\bar{D}$ from the last two terms, finishing with

$$K = A\bar{B} + \bar{A}\bar{B}\bar{C}$$

Although this takes considerable time to write down in the form of an explanation, the simplified formula can be written by an examination of the Vietch diagram, since the elements to be eliminated are so obvious.

Simplification can be accomplished in a slower way by pulling the common elements out of the original formula, always keeping in mind that the variable to be eliminated can only be one that contains both itself and its complement.

EXAMPLE

$$K = \underbrace{A\bar{B}\bar{C}D + A\bar{B}\bar{C}\bar{D}}_{} + \underbrace{A\bar{B}C\bar{D} + A\bar{B}CD}_{} + \underbrace{\bar{A}\bar{B}\bar{C}D + \bar{A}\bar{B}C\bar{D}}_{}$$

eliminating $A\bar{B}\bar{C}$ $A\bar{B}C$ $\bar{A}\bar{B}\bar{D}$ eliminating
D and $\bar{D}$ D and $\bar{D}$
eliminating $A\bar{B}$
C and $\bar{C}$

$$K = A\bar{B} + \bar{A}\bar{B}\bar{C}$$

This method may look just as easy at first glance, but it is not really as simple or as reliable and accurate as the Vietch diagram method of simplification.

We have showed the development of Vietch diagrams from Boolean expressions. Now we will attempt to move from a simple practical problem, through the development of a truth table, to the Boolean expression and finally to the Vietch diagram in an attempt at simplification.

Assume that you have a hall light with a switch at either end. Either switch should turn the light on or off. Let us assume that:

$$A = \text{switch 1 on}$$
$$\bar{A} = \text{switch 1 off}$$
$$B = \text{switch 2 on}$$
$$\bar{B} = \text{switch 2 off}$$
$$C = \text{light that glows}$$

This can be expressed: $\bar{A}B + A\bar{B} = C$ [(Switch A is off and switch B is on or A is on and B is off to get the light (C) to glow.]

Truth Tables:

Basic Reference		$\bar{A}B$ Table			$A\bar{B}$ Table			$\bar{A}B + A\bar{B}$ Table		
A	B	$\bar{A}$	B	$\bar{A}B$	A	$\bar{B}$	$A\bar{B}$	$\bar{A}B$	$A\bar{B}$	$\bar{A}B + A\bar{B}$
0	0	1	0	0	0	1	0	0	0	0
0	1	1	1	1	0	0	0	1	0	1
1	0	0	0	0	1	1	1	0	1	1
1	1	0	1	0	1	0	0	0	0	0

Now let us try a Vietch diagram to see if simplification is possible.

$\bar{A}B + A\bar{B}$

No simplification is possible.

The rule for simplifying a Vietch diagram is: If two boxes are adjacent to each other (⬜⬜ ⬜), simplification is possible. The boxes above are not adjacent, so no simplification is possible.

Vietch diagrams are often used in developing logic circuitry by simplifying the equations that constitute the circuitry. Much more can be said about these diagrams, but a detailed study belongs more properly in a course in circuit design rather than an introductory course in computer mathematics.

Boolean algebra is a tool used in symbolic logic to aid in the attempt to put logic on a mathematical foundation. Symbolic logic is discussed in Chapter 5.

REVIEW QUESTIONS

4.22. Complete a Vietch diagram for the following expression.

$$P = A\bar{B}C + \bar{A}BC + AB\bar{C} + \bar{A}\bar{B}C$$

4.23. Simplify the expression in problem 4.22.

4.24. Develop a Vietch diagram for two variables.

4.25. Mark X's in your Vietch diagram for the following expression.

$$A = AB + \bar{A}B + A\bar{B} + AB$$

Simplify the expression.

4.26. Complete a Vietch diagram for the following expression.

$$K = AB\bar{C}\bar{D} + AB\bar{C}D + ABC\bar{D} + ABCD$$

4.27. Simplify the expression in problem 4.26.

REVIEW

Review of Symbols

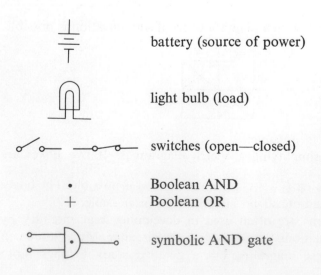

battery (source of power)

light bulb (load)

switches (open—closed)

· Boolean AND
+ Boolean OR

symbolic AND gate

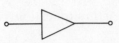 symbolic OR gate

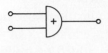 inverter

$\bar{A}$ bar over a symbol signifies the opposite of an existing state (NOT or complement)

NOT Truth Table	*AND Truth Table*	*OR Truth Table*
$\bar{0} = 1$	$0 \cdot 0 = 0$	$0 + 0 = 0$
$\bar{1} = 0$	$0 \cdot 0 = 0$	$0 + 1 = 1$
	$1 \cdot 0 = 0$	$1 + 0 = 1$
	$1 \cdot 1 = 1$	$1 + 1 = 1$

Review of Terminology

Schematic	line drawing of a circuit or part of a circuit
Source	source of power in a circuit
Load	a component of a circuit that provides a use for electrical power
Series	switches connected end-to-end

Parallel	switches connected side-by-side

AND	Boolean expression for the concept of switches in series
OR	Boolean expression for the concept of switches in parallel
Gate	a circuit component providing an output from the circuit
Enabled	this describes a gate which is conducting current
Disabled	this describes a gate which is not conducting current
Inverter	circuitry component that has the ability to change 1 to 0 and 0 to 1

Postulate	proposition that may be taken for granted
Theorem	rule that can be proven to be true
Commutative laws	the order of addition or multiplication have no effect on the result
Associative laws	when three or more items are added or multiplied, the order of addition or multplication is inconsequential
Distributive law	if two (or more) numbers are to be added, then multiplied by a third number, the result is the same if each number is first multiplied, then the products added together
Truth tables	a method used to prove Boolean theorems

SUMMARY OF CHAPTER 4

1. All variables in a Boolean equation have the values of either 0 or 1.

2. A drawing of an electrical circuit is called a *schematic*.

3. A basic electrical circuit consists of a *source*, a *switch*, and a *load*. Consider the source to be the *input* and the load to be the *output*. The switch is used to break the circuit.

4. Switches may be connected in *series* or in *parallel*, or a combination of both series and parallel.

5. No flow of current is depicted with a *zero* (0). Flow of current is depicted with a *one* (1).

6. Switches in series are represented by the Boolean AND operation.

7. Switches in parallel are represented by the Boolean OR operation.

8. An AND or an OR component in a circuit is called a *gate*. A gate is *enabled* if the output is a 1; it is *disabled* if the output is a 0.

9. A straight line over a character or symbol is the *negation* or *not* sign ($\bar{A}$). It is also written as the *prime* of a character (A').

10. The negation symbol above a character represents the *complement* of that character.

11. In a symbolic circuit, the *inverter* is used to perform the logical operation of complementation.

12. Each Boolean *theorem* (rule) has its *opposite theorem*. The principle of duality is known as De Morgan's theorem.

13. The basic AND and OR truth tables are:

AND ($\cdot$) Truth Table	OR ($+$) Truth Table
1. $0 \cdot 0 = 0$	1. $0 + 0 = 0$
2. $0 \cdot 1 = 0$	2. $0 + 1 = 1$
3. $1 \cdot 0 = 0$	3. $1 + 0 = 1$
4. $1 \cdot 1 = 1$	4. $1 + 1 = 1$

14. There are many Boolean theorems, but a few of them are particularly important:

    (a) *Commutative laws.* The order of addition or multiplication has no effect on the result.

    (b) *Associative laws.* When three or more items are to be added or multiplied, the result will be the same regardless of which sequence of addition or multiplication is taken.

    (c) *Distributive law.* If two numbers are to be added and multiplied by a third number, the result will be the same if each of the numbers to be added is multiplied by the third number, and then the individual products added.

15. All Boolean theorems may be proven to be true by developing truth tables for each side of the equation and noting that the results are identical.

16. Such truth tables are developed by working out each step of the equation with the use of the basic AND and OR truth tables.

17. Boolean terms and equations may be shown as schematics of electrical circuits or as symbolic circuits.

18. Vietch diagrams are used to pictorially represent and to simplify Boolean equations.

REVIEW QUESTIONS

4.28. Using the circuit shown below and the switches indicated, specify whether the light will glow or not glow.

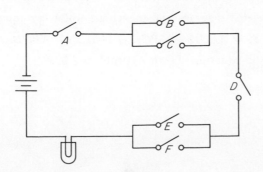

Glow   Not Glow

(a) $A = 1, B = 0, C = 1, D = 0, E = 1, F = 1$ □   □
(b) $A = 1, B = 1, C = 1, D = 1, E = 0, F = 1$ □   □
(c) $A = 1, B = 1, C = 0, D = 1, E = 0, F = 0$ □   □
(d) $A = 0, B = 1, C = 1, D = 1, E = 1, F = 1$ □   □

4.29. Write the circuit above as a Boolean expression.

4.30. What will be the output from each of the following symbolic circuits? Show the output in each phase.
Circuit 1:

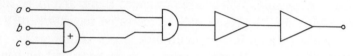

(a) $a = 1, b = 1, c = 0$ output $= □$
(b) $a = 0, b = 1, c = 1$ output $= □$
(c) $a = 0, b = 0, c = 1$ output $= □$
Circuit 2:

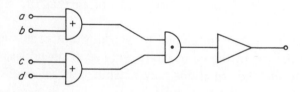

(a) $a = 0, b = 0, c = 1, d = 0$ output $= □$
(b) $a = 1, b = 1, c = 0, d = 0$ output $= □$
(c) $a = 1, b = 0, c = 1, d = 0$ output $= □$

4.31. Construct truth tables for each side of the following theorem.

$$A + (B + C) = (A + B) + C$$

4.32. Write the theorem that will be the complement of the theorem in problem 4.31 and construct truth tables for each side of the new theorem.

4.33. Draw a Vietch diagram and depict the following equation.

$$K = A\bar{B}\bar{C}D + A\bar{B}C\bar{D} + \bar{A}\bar{B}\bar{C}D + \bar{A}\bar{B}C\bar{D}$$

4.34. Simplify the equation shown in problem 4.33.

5 INTRODUCTORY LOGIC

BASIC LOGIC FORMS

Symbolic logic very closely resembles Boolean algebra, primarily because Boolean algebra is a form of logic rather than a form of mathematics. Symbolic logic is a theory and methodology of deduction and formal reasoning. It consists, in its simplest form, of one or more premises from which logical conclusions may be reached.

Conditional Statements

A conditional statement is one that gives a choice of one or more conditions from which to choose.

EXAMPLE

Premises: 1. *If* there are apples on the tree *then* Johnnie will get sick from eating too many apples.
2. There are no apples on the tree.

Conclusion: Johnnie will not get sick from eating too many apples.

Such a statement, with the form *if . . . then*, is called a *conditional*. It can easily be converted to letter symbols:

Premises:	if *p* then *q*
	not *p*
Conclusion:	not *q*

(*p* is "there are apples on the tree.")
(*q* is "Johnnie will get sick . . . .")
(Not *p* is "there are no apples . . . .")
(Not *q* is "Johnnie will not get sick . . . .")

This statement may take another form:

Premises:	if *p* then *q*
	p
Conclusion:	*q*

Using the example above, we now have:

Premises: 1. *If* there are apples on the tree *then* Johnnie will get sick from eating too many apples.
          2. There are apples on the tree.

Conclusion: Johnnie will get sick from eating too many apples.

A third form of the *if . . . then* conditional statement has the form:

Premises:	if *p* then *q*
	if *q* then *r*
Conclusion:	if *p* then *r*

This type is a combination of three statements (*p*, *q*, and *r*), two of which appear in the conclusion.

EXAMPLE

Premises: 1. *If* John steals apples *then* Bill will tell his mother.
          2. *If* Bill tells his mother *then* John will be punished.

Conclusion: If John steals apples, he will punished.

OR Statements

There are many other ways the if . . . then form can be constructed, but let us now look at a form that uses OR as the connective.

EXAMPLE

Premises: 1. Jack was born in Los Angeles *or* he was born in Chicago.

2. Jack was born in Chicago.

Conclusion: Jack was not born in Los Angeles.

This can be written symbolically:

> Premises: 1. *p* or *q*
> 2. *q*
> Conclusion: not *p*

Of course, this can be reversed:

> Premises: 1. *p* or *q*
> 2. *p*
> Conclusion: not *q*

This type of OR statement is called *exclusive OR* because either one or the other of the two initial statements may be true, but both *cannot* be true. The *inclusive OR* implies a relationship where either one *or* the other *or* both statements are true. (Recall the inclusive OR in Boolean algebra, in Chapter 4.)

In this text, the use of OR will always be in the inclusive sense unless otherwise specified.

EXAMPLE

Premises: 1. Smith has a Ford *or* he has a Pontiac.

2. He does not have a Ford.

The fact that he has (or does not have) a Ford does not preclude the possibility that he may also have a Pontiac. In the earlier example, Jack could only have been born in one of the two cities. There is no possibility that both statements could be true. This is the major difference between the exclusive and inclusive OR.

Conclusion: Smith has a Pontiac.

Symbolically, this is shown:

> Premises: *p* or *q*
> not *p*
> Conclusion: *q*

Again, this can be reversed:

Premises: p or q
           not q
Conclusion: p

AND Statements

Another form uses the connective AND, which may be shown by the following examples.

EXAMPLES

Premise: John *and* George went to the game.
Conclusion: John went to the game.
Premises: 1. John went to the game.
           2. George went to the game.
Conclusion: John *and* George went to the game.
The forms of the AND statements are as follows:

Premise: p and q	p and q	1. p
Conclusion: p	q	2. q
		Conclusion: p and q

A number of forms are drawn together in Table 5.1. Each is very simple, but it is possible to work out fairly complex problems by successive application of the forms.

Table 5.1. A Number of Simple Logic Forms

(a) if p then q (d) p or q (h) p and q
    if q then r not p ―――――――
    ―――――――   ――― p
    if p then r q
(b) if p then q (e) p or q (i) p and q
    p not q ―――――――
    ―――   ――― q
    q p
(c) if p then q (f) not (p and q)   (j) p
    not p ―――――――――――― q
    ―――   not p or not q ―――
    not q p and q
                           (g) not (p or q)
                               ―――――――――――
                               not p and not q

Note: The conclusion is below the line in each set.

Examples of each form shown in Table 5.1 follow:

(a) if p then q	If John drives too fast then he will get a ticket.
if q then r	If John gets a ticket then he will pay a fine.
if p then r	If John drives too fast then he will pay a fine.
(b) if p then q	If John gets a motorcycle then he will ride it in the desert.
p	John gets a motorcycle.
q	He will ride it in the desert.
(c) if p then q	If the bill is not paid then the electricity will be turned off.
q	The electricity is turned off.
p	The bill was not paid.
(d) p or q	Bob's new car is blue or red.
not p	It is not blue.
q	It is red.
(e) p or q	Bob's new car is blue or red.
not q	It is not red.
p	It is blue.
(f) not (p and q)	Tires may not be expensive and poorly made.
not p or not q	Tires may not be expensive or tires may not be poorly made.
(g) not (p or q)	John will not get a motorcycle or a car.
not p and not q	John will not get a motorcycle and will not get a car.
(h) p and q	There are apples and oranges in the refrigerator.
p	There are apples in the refrigerator.
(i) p and q	There are apples and oranges in the refrigerator.
q	There are oranges in the refrigerator.
(j) p	There are shirts in the closet.
q	There are pants in the closet.
p and q	There are shirts and pants in the closet.

There are many other variations of these forms, but they all fit in a similar pattern. For example, examine forms (d) and (e):

$$\text{(d) } p \text{ or } q \qquad \text{(e) } p \text{ or } q$$
$$\frac{\text{not } p}{q} \qquad\qquad \frac{\text{not } q}{p}$$

Other forms can be written:

$$p \text{ or } q \quad \text{and} \quad p \text{ or } q$$
$$\underline{p} \qquad\qquad \underline{q}$$
$$\text{not } q \qquad\quad \text{not } p$$

EXAMPLE

The office door is brown or gray.
The office door is brown.
_____

The office door is not gray.

REVIEW QUESTIONS

5.1. Work out a set of statements for each of the ten forms in Table 5.1.

5.2. What is the form of a conditional statement?

5.3. What is the difference between the inclusive and the exclusive OR?

5.4. Relate the AND type statements to switch settings as discussed in Chapter 4. Draw two switches in the AND mode.

5.5. Relate the inclusive OR type statements to switch settings as discussed in Chapter 4. Draw two switches in the inclusive OR mode.

5.6. Given the following statements.
(a) John has five marbles and two sticks of gum.
(b) An employee can retire if he reaches age 67 or if he has completed 25 years on the job.
(c) If $A > B$, then $B < C$.
Show the symbolic form and a conclusion for each of the above statements. Simplify the writing of the forms by using the following symbols:

AND   ·
OR $+$
NOT $\bar{A}$ (bar over symbol); e.g., NOT p is written $\bar{p}$

TRUTH TABLES

Truth tables are used in symbolic logic to prove the validity of simple statements. In each of the forms mentioned in the previous section, the important question was, "Is this statement true or false?" The construction of truth tables simplifies the exercise of determining the truth-falseness of a statement.

Assuming that every statement is either true or false, but not both, there are only four possibilities available for any two statements: (1) both statements are true, (2) the first statement is true and the second statement false, (3) the second statement is true and the first statement false, (4) both statements are false. In other words, there are four possible combinations of two statements, taken two at a time. This is symbolically shown in Table 5.2.

Table 5.2. Possible Combinations
for Two Statements

p	q
T	T
T	F
F	T
F	F

Using the two basic columns in Table 5.2, we can add a column titled $p \cdot q$ to get the truth table for an AND condition. This is shown in Table 5.3.

Table 5.3. AND Truth Table

$p\ q$	$p \cdot q$
$T \cdot T =$	T
$T \cdot F =$	F
$F \cdot T =$	F
$F \cdot F =$	F

The inclusive OR truth table is shown in Table 5.4.

Table 5.4. Inclusive OR Truth Table

p	q	$p + q$
$T + T =$		T
$T + F =$		T
$F + T =$		T
$F + F =$		F

You can see at a glance that for AND both statements must be true for the entire sequence to be true, and for OR, it is only false if both statements are false.

The truth table for the if . . . then condition is shown in Table 5.5. The conditional statement is considered to be false only if the *antecedent* (First premise) is true and the *consequent* (Second premise) is false. Otherwise, the conditional is considered to be true.

Table 5.5. If . . . Then Truth Table

p	q	*if p then q*
T	T	T
T	F	F
F	T	T
F	F	T

The if . . . then condition can be shown by the symbol $\supset$ ($p \supset q$, then means if p then q).

With this new truth table, let us return to the basic logic forms. Consider, for example, the form:

$$p \supset q$$
$$p$$
$$\overline{}$$
$$q$$

[This is form (b) on page 102.]

In truth table form, this would be written:

(Premise 1)	(Premise 2)	(Conclusion)
$p \supset q$	p	q
T	T	T
F	T	F
T	F	T
T	F	F

($p \supset q$)	Premise 1:	If John gets a motorcycle then he will ride it in the desert.
	Premise 2:	John gets a motorcycle.
	Conclusion:	He will ride it in the desert.

This is the (*T T T*) portion of the truth table. The other portions of the table can be checked in the same manner.

(F T F)   Premise 1 is false; therefore, even if he gets the motorcycle (premise 2: T), he will not ride it in the desert (conclusion: F).

(T F T)   Premise 1 is true, premise 2 is false, but we cannot discount the possibility (based on premise 1) that he will ride in the desert anyway (conclusion: T).

(T F F)   Premise 1 is true, but he does not get the motorcycle (premise 2: F), so he will not ride in the desert (conclusion: F)

The proper "logic" terminology for the forms studied so far is:

Conjunction	(AND)
Disjunction	(OR)
Negation	(NOT)
Implication	(IF ... THEN)

The results of all three truth tables may be placed into a single table, as shown in Table 5.6.

Table 5.6. Truth Tables for $\cdot$, $+$, $\supset$

p	q	$p \cdot q$	$p + q$	$p \supset q$
T	T	T	T	T
T	F	F	T	F
F	T	F	T	T
F	F	F	F	T

The Biconditional Form

A statement (consisting of two factors) that is true if both factors are true, true if both factors are false, and false otherwise, is called *biconditional*. It is expressed in English with "if and only if."

EXAMPLE

The hunter will fire his gun if and only if the lion charges.
The "if and only if" statement implies equivalence. p if and only if q may be expressed as:

$$p \text{ implies } q \text{ AND } q \text{ implies } p \dots$$

(The symbol for "implies" is $\rightarrow$.) therefore, if $p \rightarrow q$ AND $q \rightarrow p$, this can be abbreviated to $p \rightarrow q$, or p is equivalent to $q(p \equiv q)$.

This type of statement is expressed symbolically with $\equiv$, as $A \equiv B$ (assuming A is "the hunter will fire" and B is "the lion charges"). A is true if and only if B occurs.

The truth table for this form is shown in Table 5.7.

Table 5.7. Biconditional Truth Table

$p \equiv q$ Truth Table
$T \equiv T = T$
$T \equiv F = F$
$F \equiv T = F$
$F \equiv F = T$

Negating the above example is the only other way to reach the conclusion of "true." The hunter will *not* fire his gun if and only if the lion does *not* charge. The truth table for the NOT condition is shown in Table 5.8. Obviously, it is exactly the opposite of the $p \equiv q$ truth table.

Table 5.8. Biconditional NOT Truth Table

$(\overline{p \equiv q})$ Truth Table
$T \equiv T = F$
$T \equiv F = T$
$F \equiv T = T$
$F \equiv F = F$

The truth conditions in the $(\overline{p \equiv q})$ truth table can be interpreted as the exclusive OR condition; the "true" conclusion is only reached when one factor is true and the other factor is false.

EXAMPLE

$(\overline{A \equiv B})$ (T) A—the hunter will shoot
            (T) B—the lion will charge
            (F) A—the hunter will not shoot
            (F) B—the lion does not charge

By writing out each statement as it relates to the truth table above, it is easy to check the accuracy of the truth table:

1. It is not true that the hunter will shoot if the lion charges. False (because he will shoot under that condition).

2. It is not true that the hunter will shoot if the lion does not charge. True.

3. It is not true that the hunter will not shoot if the lion charges. True.

4. It is not true that the hunter will not shoot if the lion does not charge. False.

For simple statements, such as the ones used on the previous pages, it is easy to explain the truth or falsity of a statement in English, but for more complex forms, the truth tables are quite essential.

REVIEW QUESTIONS

5.7. What is the purpose of a truth table?

5.8. Show all the available possibilities for two statements, assuming that every statement is either true or false. Use the letters "p" and "q" to designate the two statements.

5.9. What symbol is used to designate the if . . . then condition?

5.10. In a conditional statement:
(a) What is the first premise called?
(b) What is the second premise called?

5.11. What condition must prevail for a simple conditional statement to be false?

5.12. Show the truth table for:
(a) $p \cdot q$
(b) $p + q$
(c) $p \supset q$

CONDITIONAL STATEMENTS RELATED TO COBOL

COBOL (COmmon Business Oriented Language) is the computer language most commonly used in business systems. Conditional expressions are used extensively in the writing of COBOL programs.

In COBOL, conditional expressions are separated into two major types: *simple* and *compound*. If the conditional expression contains just one condition, it is called a *simple* conditional expression. If the expression contains two or more conditions, it is a *compound* expression.

The logical operators NOT, AND, OR are used in conditional expressions. NOT is used in simple conditional expressions; AND, OR is used in compound conditional expressions.

IF *A* IS EQUAL TO B GO TO TOTALS.   (simple—just one condition)
IF *A* IS NOT EQUAL TO B GO TO TOTALS.   (simple)
IF *A* AND *B* = *C* GO TO TOTALS.   (compound)
IF *A* OR *B* = *C* GO TO TOTALS.   (compound)

In the two compound statements above, the connective THEN is not written out, but is implied. They could have been written:

IF *A* AND *B* = *C* THEN GO TO TOTALS.
IF *A* OR *B* = *C* THEN GO TO TOTALS.

Just as before OR means either one or both conditions must be true and AND means both conditions must be true.

Implications and Equivalences

Common factors may be implied instead of being repeated over and over, if the compound expression consists of a series of simple expressions. If two expressions have identical meanings, they are said to be *equivalent*. In Chapter 1, we defined equivalence as two sets having the same number of elements (or a one-to-one relationship existing between the sets). Equivalence in the present context refers to two statements, said in different ways, but meaning the same thing.

EXAMPLES

Simple Conditional Expressions

Assume, in the first two statements, that FIELD–A will always contain either P–N 127 or +999:

IF FIELD–A IS EQUAL TO 'P–N 127' GO TO REPEAT.
IF FIELD–A IS NOT EQUAL TO '+999' GO TO REPEAT.

These two statements are equivalent because they both accomplish the same purpose.

IF FIELD–B IS NOT GREATER THAN FIELD–C GO TO RESTART.

IF FIELD–B IS LESS THAN FIELD–C GO TO RESTART.

(These last two expressions are not equivalent. If FIELD–B is not greater, it may be either = or < and the = is not covered.)

Compound Conditional Expressions

IF FIELD–A IS EQUAL TO FIELD–B OR FIELD–A IS NOT LESS THAN FIELD–C GO TO AREA–1.
IF FIELD–A IS EQUAL TO FIELD–B OR NOT LESS THAN FIELD–C GO TO AREA–1.

Implied Subjects

The last two statements in the examples above are equivalent. The second statement uses an *implied* subject. The rule is that *common factors* in a compound expression may be implied if the compound expression consists of a group of simple expressions. In the examples above, the second use of FIELD–A was omitted in the actual statement, but it is implied and will be used as a source of comparison by the computer.

EXAMPLE

IF AREA–1 IS GREATER THAN 100 AND AREA–1 IS LESS THAN 200 . . . .

This could be written:

IF AREA–1 IS GREATER THAN 100 AND LESS THAN 200 . . . .

When the subject is implied, the subject used is the first subject to the left which is explicitly stated.

EXAMPLES

1. IF A = B OR IS LESS THAN C AND GREATER THAN D GO TO ROUTINE–1.
2. IF A LESS THAN B OR LESS THAN C OR LESS THAN D OR E LESS THAN F GO TO ROUTINE–2.

This last example would be interpreted as:

$$A < B \quad \text{or} \quad A < C \quad \text{or} \quad A < D \quad \text{or} \quad E < F$$

Notice that A is the first stated subject, through D, then E is the stated subject for F.

Implied Operators

If operators (arithmetic: $+$, $-$, etc.; logical: AND, OR, NOT; relational: $>$, $<$, $=$) have the same type of relationship as is mentioned above for subjects, they may also be implied instead of being repeated.

EXAMPLES

1. IF PRICE IS LESS THAN COST OR PRICE IS LESS THAN MARGIN . . . . This could be written: IF PRICE IS LESS THAN COST OR MARGIN . . . . In this case, both the *subject* (PRICE) and the *operator* (IS LESS THAN) are implied in the second half of the expression.

2. IF A LESS THAN B OR C OR D OR E LESS THAN F GO TO ROUTINE–2. This statement is identical to example 2 in the previous example above. The unnecessary repetition of both the subject (A) and the operator (LESS THAN) has been omitted.

Conjunctions and Disjunctions

Consider the following statement:

IF A AND B = C GO TO ROUTINE–1.

This is a simple "if . . . then" statement with a logical AND connective.

$$p \supset q \text{ (if } p \text{ then } q)$$

p is $A \cdot B = C$ q is GO TO ROUTINE–1

A logical AND is often referred to as a *conjunction*.

Now consider this statement:

IF A OR B = C GO TO ROUTINE–1.

This is also a simple "if . . . then" statement with a logical OR connective.

$$p \supset q \text{ (if } p \text{ then } q)$$

p is $A + B = C$ q is GO TO ROUTINE–1

A compound logical OR is often called a *disjunction*. A statement is considered to be *compound* if there are two or more connectives:

IF A OR B OR C = D OR E = F GO TO ROUTINE 1.

Most COBOL statements that require decisions to be made can be related to logic forms, thus simplifying what might otherwise seem to be extremely complex statements.

REVIEW QUESTIONS

5.13. We want to find out if the data in AREA-1 is a positive number. If it is other than positive, we want to go to a SPECIAL-ROUTINE. Write the expression to accomplish this action.

5.14. The on condition of switch 1 is named PRINT. Write an expression that will go the PRINT-ROUTINE if the switch is on and to TAPE-ROUTINE if the switch is off. Write this with two simple statements.

5.15. Write the following expression in a simpler manner. IF A GREATER THAN B AND A = C OR D LESS THAN E OR D LESS THAN F GO TO ROUTINE-1.

5.16. Name the three relational operators.

5.17. Name the three logical operators.

5.18. A relational operator can be implied only if the subject is also implied. Is this statement true or false?

5.19. What is the meaning of the term equivalence?

5.20. What is a conjunction?

5.21. What is a disjunction?

FLOW CHARTING

The purpose of studying systematic techniques of logic is to provide a method for problem solving. This is important because the biggest part of a computer programmer's job is the study and solution of problems.

One of the important methods used by programmers is the technique of flow charting. Flow charts are developed to aid in defining a problem, developing the logic, determining that no areas are left uncovered, and to help in the writing of the program. Flow charts also serve as effective documentation of what the program is accomplishing.

Symbology

There are quite a number of symbols connected with the area of flow charting. For the purpose of this chapter, only a few will be developed.

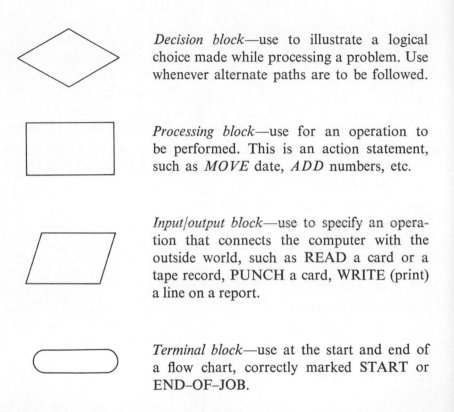

Decision block—use to illustrate a logical choice made while processing a problem. Use whenever alternate paths are to be followed.

Processing block—use for an operation to be performed. This is an action statement, such as *MOVE* date, *ADD* numbers, etc.

Input/output block—use to specify an operation that connects the computer with the outside world, such as READ a card or a tape record, PUNCH a card, WRITE (print) a line on a report.

Terminal block—use at the start and end of a flow chart, correctly marked START or END–OF–JOB.

EXAMPLE

Draw a simple flow chart to depict the following action: Input to the computer is a deck of punched cards. Each input card contains four *fields* (an area of the card designed to hold a particular piece of information). Each field contains a number. The problem is to add fields 1 and 2, subtract field 3 from field 4, and punch output cards containing the sum and difference only.

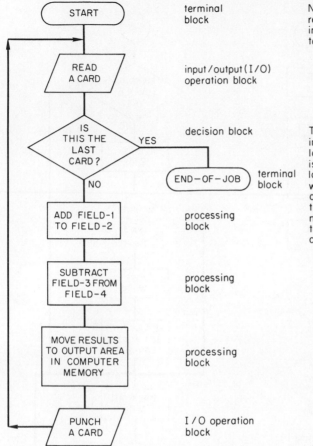

terminal block	Notice how simple it is to read the flow chart and immediately grasp what is to happen in the program.
input/output(I/O) operation block	
decision block	Test for a special last card in the card deck. If it is the last card (YES), the process is finished. If it is not the last card (NO), the process works its way through one card and then goes back to the beginning to read the next card, continuing around the loop until eventually the last card is reached.
terminal block	
processing block	
processing block	
processing block	
I/O operation block	

Any process that can be accomplished can be flow charted. Figure 5.1 is a humorous little flow chart that demonstrates this fact.

Decision Blocks

The most important blocks in any flow chart are the decision blocks. The logic of a process is nearly always expressed with this symbol. A great deal of symbology is related to the decision block, primarily because this is the block that causes the program to make a choice between two or more paths to follow, depending on the condition being tested.

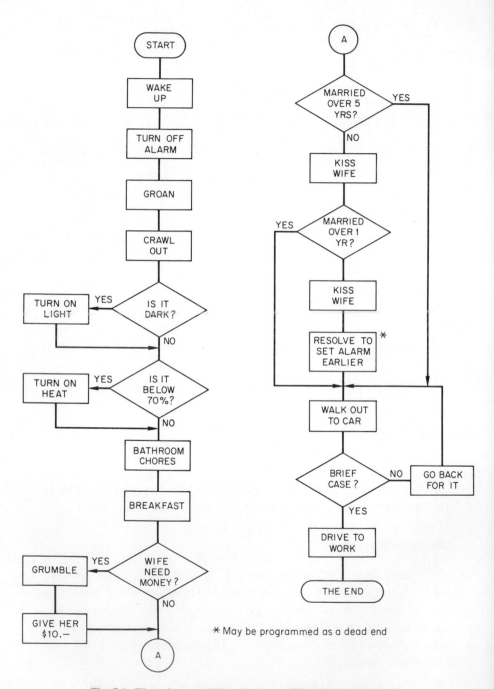

Fig. 5.1. Flow chart on "How To Get to Work in the Morning."

A few of the possible conditions and some of the associated symbology are shown in the list below and in Fig. 5.2.

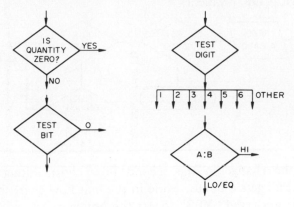

Fig. 5.2. Decision symbols.

Statement	Symbol
compare A with B (where B is the constant value)	$A:B$
A is greater than B	$A > B$
A is less than B	$A < B$
A is equal to B	$A = B$
A is not equal to B	$A \neq B$
A is less than, or equal to, B (not greater)	$A \leqslant B$
A is greater than, or equal to, B (not less)	$A \geqslant B$
compare indicator settings	HI LO EQ
check indicator settings	ON OFF

Flow Charting Logical AND and OR

Consider the following problem: A company wishes to give a bonus to all employees who are 55 years old and who have worked for the company for 15 years. This is a simple logical AND problem: Chart 1:

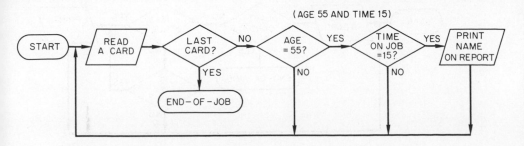

Now the same problem, slightly changed: A bonus will be given to all employees who are 55 years old or who have worked for 15 years. This is a logical OR problem:

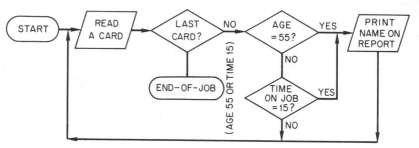

Notice the change in the second flow chart. Either question answered "YES" gets a bonus, while in the first flow chart, both questions must be answered "YES" to get the bonus.

The two examples on the previous page can be reduced to simple true and false statements by giving each factor a symbol.

$$p = \text{age} = 55$$
$$q = \text{time on job} = 15$$
$$X = \text{print name on report}$$

Chart 1: $\quad X = p \cdot q \qquad \bar{X} = \bar{p} + \bar{q}$

Chart 2: $\quad X = p + q \qquad \bar{X} = \bar{p} \cdot \bar{q}$

The flow charts could be redrawn as follows:

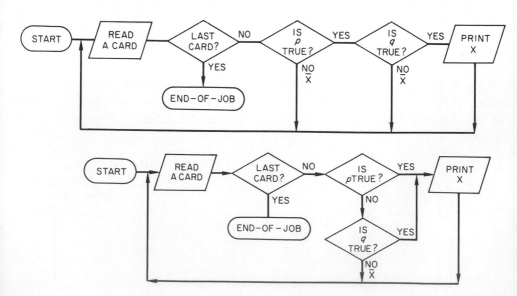

To drive this point home, we will try another pair of problems with more variables. Notice that the basic method does not change.

A bonus will be given only if the employee meets all of these requirements:

$$\text{Age 55 or over} \qquad (p)$$
$$\text{Time on job 15 years} \quad (q)$$
$$\text{Salary below 5000} \qquad (r)$$
$$\text{Male employee only} \qquad (s)$$

This can be expressed:

$$\text{Bonus} = p \cdot q \cdot r \cdot s$$

Flow chart:

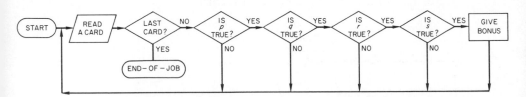

The problem can be changed to specify that a bonus will be given if any one of the above conditions is true. The expression will be:

$$\text{Bonus} = p + q + r + s$$

Flow chart:

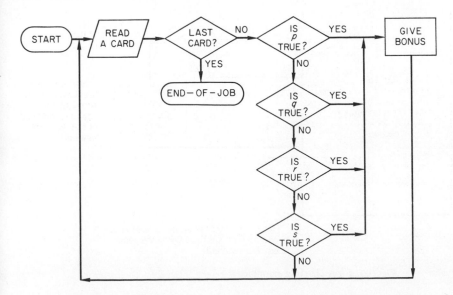

EXAMPLE

Flow Charting Problem Statement

The annual inventory for a company is punched into cards with the following format:

Description

Stock number

Quantity counted

Order code

If the quantity counted is 0 or if the order code is X, only the stock number is to be punched on a card. Otherwise, all positions of information are to be punched on a card.

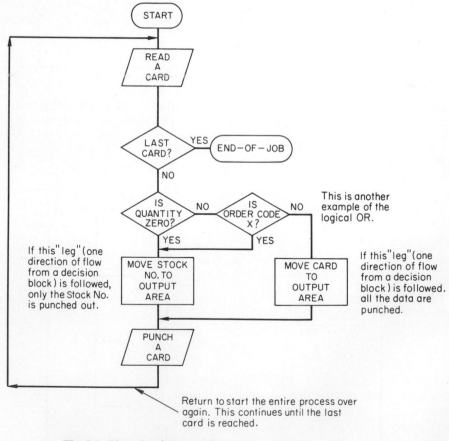

Fig. 5.3. Flow charting example.

Steps in working out the flow chart:

1. Terminal (start).
2. Read a card.
3. Test for special last card (decision block—"yes" or "no").
(a) If "yes," halt the program (terminal block).
(b) If "no," continue program.
4. Is quantity zero? (Decision block—"yes" or "no.")
(a) If "yes," move stock number to output area.
(b) If "no", is order code "X"? (Decision block—"yes" or "no.")
    (1) If "yes," move stock number to output area.
    (2) If "no," move card to output area.
5. Punch a card (card will contain contents of output area).
6. Go back to "start" in order to repeat the entire process.

A flow chart for this process is shown in Fig. 5.3. It is not really complete and, of course, the problem itself is very insignificant, but it should give the student an idea of how a problem is analyzed and flow charted.

REVIEW

Review of Symbols

⊃   if . . . then . . .
≡   if and only if (biconditional) symbol for equivalence

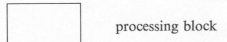

decision block

processing block

input/output (I/O) block

terminal block

:	compare
$\neq$	unequal
$\geqslant$	greater than or equal to
$\leqslant$	less than or equal to
$\rightarrow$	implies $(A \rightarrow B)$

Review of Terminology

Conditional — a statement that gives one or more conditions from which to choose

Exclusive OR — one or the other of the conditions are true, but both cannot be true

Antecedent — first premise

Consequent — second premise

Biconditional — statement that is true only if both conditions are true or both conditions are false

COBOL — COmmon Business Oriented Language

Compound — conditional expression containing two or more conditions

Simple — conditional expression containing just one condition

Equivalent — two or more expressions with identical meaning

Implied — factors may be implied instead of being repeated; both the subject and operator may be implied under the proper conditions

Conjunction — logical AND

Disjunction — compound logical OR

Field — an area designed to hold a particular piece of information

REVIEW QUESTIONS

5.22. Identify the following flow charting blocks by name and briefly describe the function of each.

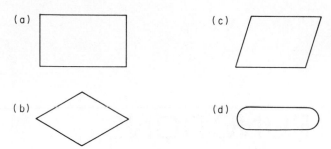

(a)

(b)

(c)

(d)

5.23. Flow chart the following logic problem: Given a deck of cards, each card has only two fields (titled *A* and *B*), which contain numbers that are never equal. For each card, determine which field contains the largest number and punch an output card consisting of the amount contained in that field.

5.24. Now for a more difficult problem, based on problem 5.23 above. Given three fields (titled *A*, *B*, and *C*) containing numbers that are never equal. Determine which field is the largest. When that is determined, the numbers in the other two fields are added together, a line is printed (written) on a report and the program continues until all cards are read.

6 FUNCTIONS AND EQUATIONS

TERMINOLOGY OF FUNCTIONS

The study of functions is basic to the study of mathematics and it aids the student in the development of skill in algebra. It is usually interesting even to nonmathematically-oriented students because a good foundation in this topic helps in the study of nearly every science.

A quick review of Chapter 1 may be of advantage at this time because the study of functions deals with sets. A function is essentially the relationship between sets.

Function can be defined as the relationship between two sets, where each element of the first set is related to just one element of the second set.

EXAMPLE

$$A = \{a, b, c\}$$
$$\downarrow \quad \downarrow \quad \downarrow$$
$$B = \{1, 2, 3\}$$

Set A is called the *domain* of the function. All elements of set B that are assigned to set A in this one-to-one relationship are called the *range* of the function.

Quite often, both the domain and the range of a function are sets of numbers, but this is not a requirement of functions. They may refer to people, objects, production schedules, school grades, accounting reports, etc.

A function may also be described as the *mapping* of the elements of its domain onto the elements of its range. In the example above, it is correct to say that a is mapped onto 1, b is mapped onto 2, c is mapped onto 3. Another way of saying the same thing is: The *image* of a is 1, the *image* of b is 2, and the *image* of c is 3.

To define any function, you must know two things: (1) the domain of the function and (2) the rule to obtain the image of an element of the domain.

EXAMPLE

$A = \{$MON, TUES, WED, THUR, FRI, SAT, SUN$\}$
$B = \{1, 2, 3, 4, 5, 6, 7\}$

Let the images of MON, TUES, WED, be 2, 4, 6 and the images for the other days be the odd numbers starting with 1 for THUR. The map of this function would be:

MON	2
TUES	4
WED	6
THUR	1
FRI	3
SAT	5
SUN	7

GRAPHS OF FUNCTIONS

A simple line graph of the function described in the example above can be drawn by letting the x axis represent the domain and the y axis represent the range. This is shown in Fig. 6.1.

The following graph has no particular meaning, but it does show a method of presenting a function visually. Remember the most important rule for functions is that every element of a domain must have one

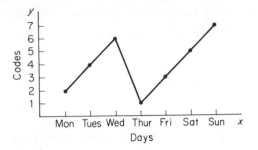

Fig. 6.1. Simple line graph of a function.

image in the range and that no element of a domain may have more than one image in the range.

In the line graph above, the range is shown as a set of points along the y axis and the domain is shown as a set of points along the x axis.

EXAMPLE

The domain is a 12-hour period, each element being one hour. The range is the size of the surf.

Map	
Hour	*Feet*
1	3
2	$3\frac{1}{2}$
3	3
4	$2\frac{1}{2}$
5	2
6	$1\frac{1}{2}$
7	1
8	2
9	$2\frac{1}{2}$
10	3
11	$3\frac{1}{2}$
12	4

The line graph for this example is shown in Fig. 6.2.

The graph in Fig. 6.1 depicts a *discontinuous* function because it consists of specific, immovable points on the graph. The graph in Fig. 6.2 depicts a *continuous* function because additional measurements could be taken at any point along the time continuum, thus forming a continuous line.

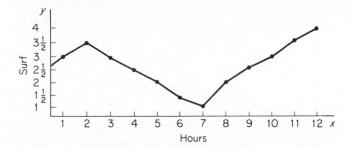

Fig. 6.2. Continuous function line graph.

REVIEW QUESTIONS

6.1. What is a function?

6.2. Consider the following two sets.

$$A = \{1, 2, 3, 4, 5\}$$
$$B = \{6, 7, 8, 9, 10\}$$

(a) Which is the domain?
(b) Which is the range?
(c) Map the function.
(d) Is this a continuous or discontinuous function? Why?
(e) Draw a line graph of the function with the domain on the y axis.

6.3. Could the following sets be sets of a function?

$$A = \{a, b, c, d\}$$
$$B = \{1, 2, 3, 4, 5\}$$

Various Types of Graphs

Problem 6.2 asked for the drawing of a line graph with the domain on the y axis. There is no law that says this cannot be done, but it is more common to plot the domain on the x axis.

It is possible to plot two, three, or even several functions on a single graph. Figure 6.3 is an example of three functions plotted on a single graph.

This type of graph requires two factors: (1) the domain must be the same for all of the functions being graphed and (2) the range must be in identical units, although there is no restriction on the images

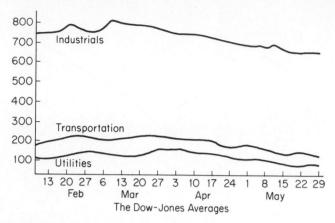

Fig. 6.3. Three functions plotted on one graph.

of the various functions. The images may be close, overlapping, or widely separated.

The rule for a function may be in algebraic terms. In this event, the symbology used is $f(x)$, which means "the function of x." Of course it could be $f(y), f(A), f(G)$, etc. A graph of this type of function is shown in Fig. 6.4.

EXAMPLE

$$x = \{-3, -1, 1, 3, 4\} \qquad \text{(This specifies the domain.)}$$

The rule will be that each element will be mapped onto $2x$. Now the total function can be written: $f(x) = 2x$ (two times x).

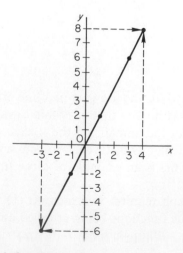

Fig. 6.4. Algebraic function graph.

Map

x	$f(x)$
-3	-6
-1	-2
1	2
3	6
4	8

The intersection of the x and y axes is designated as zero. x is then marked off in equal intervals, with the left side being minus and the right side being plus. The same procedure is accomplished along the y axis. Then it is a simple matter to lay out the graph with the aid of the map.

A function can also be graphed with a bar-graph. This is particularly good for discontinuous functions. The domain and range are laid out in the same manner as before, but instead of simply placing points on the graph and drawing connecting lines, a bar is drawn to indicate the distance along the range from the basic x-axis line.

EXAMPLE

$$x = \{M, T, W, Th, F\}$$

The range is the amount of time it takes to get to work.

Map

x	$f(x)$
M	30 min.
T	40 min.
W	25 min.
Th	30 min.
F	1 hour

Days

REVIEW QUESTIONS

6.4. Draw a graph of the following function.

$$x = \{-2, 1, 2, 3\}$$

Rule: $f(x) = 3x - 1$.

6.5. Convert Fig. 6.2 into a bar graph.

6.6. The domain of the function f is the set of real numbers.

$$f(x) = 3x + 4$$

Solve the following.

(a) $f(1) =$ (In this case, the whole number "1" replaces x in the formula.)

(b) $f(-1) =$

(c) $f(a) =$

(d) $f(a + 3) =$

This type of computation has not been shown in the chapter, but it should be solvable. Remember the rules for algebraic arithmetic discussed, in Chapter 3.

6.7. In problem 6.6 above, what element of f has the image of 28? The starting formula will be: $f(x) = 3x + 4 = 28$. $x = ?$

Rules for Functions

A function is determined by the rule that each element of the first set (domain) is related to *just one* element of the second set (range). It is permitted for an element of the second set to be related to more than one element of the first set, but the reverse is not permitted.

EXAMPLES

1. $A = \{1, 2, 3\}$
   $B = \{4, 5, 6, 7\}$

Map	
A	B
1	4
2	4
3	5

This is a function since the rule is met.

2. Same sets as above.

Map

A	*B*	
1	4 ⎫	does not comply
1	5 ⎭	with the rule
2	6	
3	7	

This is not a function because the rule is violated insofar as element 1 is assigned to both 4 and 5. There is nothing wrong with the relationship shown here, but it is not a function.

3. $A = \{1, 3, 5, 7\}$
   $B = \{2, 4, 6, 8\}$
   Rule: $A = B + 2$

Map

A	*B*
1	4
3	6
5	8
7	10

This conforms to all requirements for a function.

To show that one element of the first set cannot be logically related to more than one element of the second set, consider this simple example:

Set *A* consists of the students in a history class.
Set *B* consists of grades on one particular test.

A partial map could be:

A	*B*
John	90
Jim	82
Joan	88
Mary	79
Tom	86

It wouldn't make much sense to give John (an element in set *A*) two different grades for the same test:

John 90

John 72

REVIEW QUESTIONS

6.8. Do the following sets and rule specify a function?

Set $A =$ positive whole numbers

Set $B =$ positive whole numbers

Rule: $A = B - 3$

6.9. Does the following map specify a function?

Map	
x	$f(x)$
a	1
b	2
b	3
c	4
c	5

6.10. Does the following map specify a function?

Map	
x	$f(x)$
1	a
2	a
3	a
4	a
5	b
6	b
7	b

6.11. Do the following sets and rule specify a function?

$x = \{a, b, c, d, e\}$

$y = \{f, g, h, i, j\}$

Rule: $x = 3y + 2$

6.12. Using the following sets, draw a map and a graph of the function and state the function as an equation.

$$x = \{0, 1, 2, 3, 4, 5\}$$
$$y = \{0, 3, 6, 9, 12, 15\}$$

CARTESIAN COORDINATES

The graph shown in Fig. 6.4 is called a *Cartesian coordinate graph*. The basic format of the graph is shown in Fig. 6.5. Note that the domain is along the horizontal axis and the range is along the vertical axis. The zero point is called the *point of origin*.

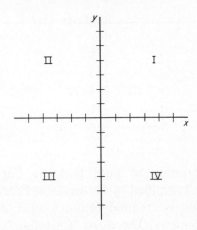

Fig. 6.5. Basic Cartesian coordinate graph.

When the domain is not limited to a few elements, but is expressed as an algebraic equation with all real numbers included in the domain, it is possible to have both positive and negative numbers and the axes can be extended to any desired length.

The axes intersect at right angles, forming four rectangles which are called *quadrants* (marked with Roman numbers in Fig. 6.5).

Any point on the graph is located by its coordinates (x and y) and may be designated by the quadrant where it is located.

EXAMPLE

$$x = 2y$$

One point could be $x = 3$ and $y = 6$; another point could be $x = -3$ and $y = -6$.

The coordinates would be written (3, 6) and (−3, −6), where the first number is always the domain and the second number is the range. (3, 6) is located in quadrant I; (−3, −6) is located in quadrant III.

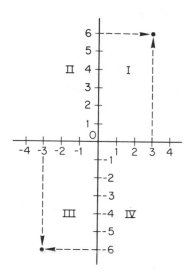

It may be interesting to note that the Cartesian rectangular relationships were first specified by Descartes, a French soldier, scholar, and philosopher, who is credited (among other achievements) with inventing analytic geometry. The word "Cartesian" is derived from the Latin translation of the name Descartes.

Whenever a function is expressed as an equation, the domain is automatically implied to be real numbers unless otherwise specified. The range is usually restricted by the equation itself.

EXAMPLE

Take a simple equation, such as the following:

$$x(x - 1) = x^2 - x$$

Now consider each side of the equation to be a set:

$$A = x(x - 1) \qquad B = x^2 - x$$

The original equation says that these two sets are equal. To prove that they are equal, simply substitute any real number for x.

$$A = 9(9 - 1) \qquad B = 9^2 - 9$$
$$= 72 \qquad\qquad = 72$$

Linear Functions

A function is called *linear* if a graph of the function forms a straight line.

EXAMPLES

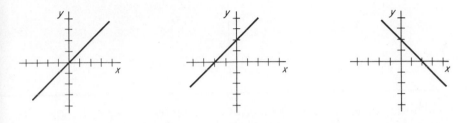

A linear function is not restricted to the rectangular form, which includes all four quadrants. If no negative numbers are involved, the form can be a single right angle with the point of origin at the junction of the two axes, as shown in Fig. 6.6. The straight line connecting the points on the graph gives it the name "linear."

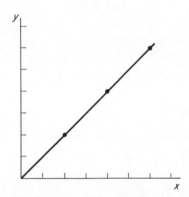

Fig. 6.6. One form of a linear function graph.

Notice, in the rectangular form, that three of the quadrants are used only when at least one of the elements is negative. See Fig. 6.7.

Only quadrant I contains two positive elements; therefore, if there are no negative elements, the other three quadrants are not needed.

There are many applications of linear functions in real life. In working with these cases, analysts and programmers convert verbal

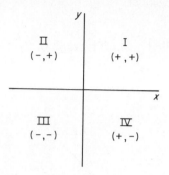

Fig. 6.7. Negative elements in quadrants.

situations into functional equations, making them easy to handle with computers.

EXAMPLE

Determine the wages for an individual working a specified number of hours for $3.00 per hour. The function to represent this is:

$$y = 3.00x$$

where

$x =$ number of hours worked
$y =$ wages ($3.00 times number of hours worked)

The graph for the above linear function would be:

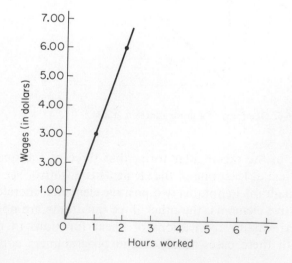

REVIEW QUESTIONS

6.13. What is the point of origin on a graph?

6.14. Into what quadrant would each of the following coordinates fall?
  (a) $(3, -3)$ (d) $(3, 3)$
  (b) $(-3, 3)$ (e) $(-3, 6)$
  (c) $(-3, -3)$ (f) $(-6, -3)$

6.15. From the following graphs, prepare maps and equations to fit each case.

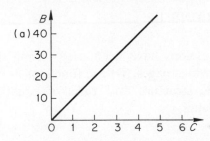

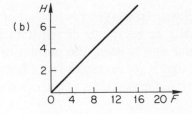

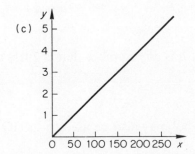

6.16. Retirement in a particular company is based on the following requirements.

Age 65 and 5 years service

Age 60 and 20 years service
Age 55 and 30 years service

A search of the personnel file revealed the following eligibles.

two 55—30 combination
three 60—20 combination
one 65— 5 combination

(a) Draw a graph of this function.
(b) Is this a linear function?
(c) Is this a continuous or discontinuous function?

FORM OF EQUATIONS

A few primitive equations have been used during the discussion of functions on the previous pages. To use functions and equations more effectively, the basic structure and rules of equations need to be examined in more detail.

Equivalence

Statements that are written in symbols instead of words are called *sentences* and it is usually necessary to decide if such statements are true or false. If the equal sign is used, the portion of the sentence on either side of the equal sign is presumed to stand for the same number or quantity. In other words, the two sides of the equation are *equivalent*.

EXAMPLES

$$5 + 9 = 14$$
$$\tfrac{8}{4} = 2$$
$$a = b - 2 \text{ (if } a \text{ is any number that is 2 less than } b)$$

Inequalities

The not equal ($\neq$) symbol may also be used to specify true statements.

EXAMPLES

$$3 \neq 5$$
$$3 + 6 - 2 \neq 10$$

These sentences specify that the left side of the sentence does not equal the right side of the sentence.

The symbols for greater than ($>$) and less than ($<$) may also be used in sentences constituting true statements.

EXAMPLES

$$3 > 1 \text{ (3 is greater than 1)}$$
$$14 < 17 \text{ (14 is less than 17)}$$
$$17 + 2 - 4 > 4 \text{ (17 + 2 - 4 is greater than 4)}$$
$$5 < 4 + 6 - 3 \text{ (5 is less than 4 + 6 - 3)}$$

The examples above represent *inequalities*.

An *open* sentence is one in which one of the elements is missing.

EXAMPLES

1. $? + 3 = 8$

The question mark is usually replaced by a symbol: $x + 3 = 8$.

2. $4 - x = 2$

3. $x > 4$

4. $x < 10$

In the first two examples, only one specific number can replace the question mark (x), but in examples 3 and 4, several numbers will fit into place. Any number 5 or greater is greater than 4 and any number 9 or less is less than 10. In example 4, it would be correct to say that any number in the *set* of real numbers from 0 through 9 (also all negative numbers) could replace the unknown quantity $\{0, 1, 2, 3, 4, 5, 6, 7, 8, 9\}$. This is called the *replacement set* and if a symbol is placed into a statement of inequality ($a < 10$), it is called a *variable*.

A variable is so called because any one of the elements of the replacement set can be substituted for it in the equation. The set in question is the *domain* of that particular variable.

REVIEW QUESTIONS

6.17. What is the definition of a sentence?

6.18. What is the symbol for equivalence?

6.19. What symbols can be used to show inequalities?

6.20. What is the definition of an open sentence?

6.21. Are the following sentences true or false?

(a) $8 + (-2) = 6$ (e) $9 \neq 4 + (-5)$

(b) $-3 + 4 = 7$ (f) $6 + 3 > 4 + 6 - 1$

(c) $-4 = -3 + (-1)$ (g) $12 < 6 + 8 - (-2)$

(d) $-10 = -4 - (+6)$ (h) $(4)(2) \neq 8$

6.22. What is the replacement set for the following statements?

(a) $x < 6$ (b) $x > 6$

6.23. Replace the variable in problem 6.22(a) with the second element in its domain.

6.24. Place the proper symbol ($=, \neq, >, <$) between the following expressions.

(a) $12 \quad 9$

(b) $4x + 2 \quad y^2 - 4$

(c) $1 + (-4) \quad -7$

(d) $6 + 3 \quad 4 + 5$

SOLVING EQUATIONS

The truth of open sentences cannot be determined without replacing the symbols with numbers. Until that is done, there is no way to decide whether a sentence is true or false.

EXAMPLES

$$x + 2 - y = 16$$
$$2x - 3 = y - 1$$
$$x^2 + 3 = 2x + 3$$

Sometimes it is possible to discover the number that should replace the unknown element in the equation. When that is possible, the equation has been *solved*.

EXAMPLE

$$x^2 + 3 = 2x + 3$$

The only numbers that will prove this equation to be true are 2 or 0. Replacing x by 2 gives:

$$2^2 + 3 = 2(2) + 3$$
$$7 = 7$$

The number that replaces the symbol (2, in this case) is called the *root* of the equation. The question is how to find the root?

Changing the Form of Equations

Equations can often be solved by changing the form and using addition or multiplication in arriving at the correct solutions. The basic method of these techniques is quite simple and straightforward.

Addition

In some very simple equations, the solution is obvious, but in other equations, it takes much more thought.

EXAMPLES

1. $5 + x = 8$. The root of this equation is obviously 3.
2. $x - 4 = 16$. The root is 20, of course.

These problems are so simple that many times the method used to solve them is not even consciously thought out. The solutions become even more obvious if the equations were changed as follows:

$$8 - 5 = x$$
$$16 + 4 = x$$

One way of solving equations is to add the *inverse* of the known element to the other side of the equation. This is exactly how the examples above were solved, but unconsciously. A more careful examination will show how it is accomplished.

EXAMPLES

1. $x + 4 = 20$

Inverse of $+4$: $\boxed{-4} \to -4$

$$x = 16$$

Proof: $16 + 4 = 20$

2. $x - 3 = 12$

   Inverse of -3: $\boxed{+3} \longrightarrow \dfrac{+3}{}$
                          $x = 15$

   Proof: $15 - 3 = 12$

3. $14 = x - 5$
   $\dfrac{+5 \longleftarrow \boxed{+5}}{19 = x}$ inverse of -5
   Proof: $14 = 19 - 5$

Changing the form of the equation in this manner for equations using addition and subtraction is called the *additive inverse* method.

Multiplication

Examine a simple multiplication example to see how it is solved.

   EXAMPLE

$$4x = 20$$

4 times $x = 20$; solve for x.

$$4\,\overline{|\,20} = 5 \qquad x = 5$$

Proof: $4(5) = 20$

What really happened in this solution was that both sides of the equation were multiplied by the inverse of 4:

$$\tfrac{1}{4}(4x) = \tfrac{1}{4}(20)$$
$$x = 5$$

The *multiplicative inverse* method is also used for division, when fractions are involved.

   EXAMPLE

$$\frac{x}{2} = 5$$

This could be written:

$$\tfrac{1}{2}x = 5$$

The simple method of solution is:

$$2 \times 5 = 10$$
$$x = 10$$

Fractions are eliminated by using the multiplicative inverse of $\frac{1}{2}$, which is 2.

$$2\left(\frac{x}{2}\right) = 2(5)$$

The methods discussed on the previous pages will work equally well if more than one term in an equation is unknown. A combination of the additive and multiplicative inverse methods may need to be used.

EXAMPLES

1. $6x = 20 - 4x$

$+4x \longleftarrow \boxed{+4x}$ inverse of $-4x$

$10x = 20$

$10\overline{\smash{)}20} = 2 \qquad x = 2$

Proof: $6(2) = 20 - 4(2)$
$\qquad\qquad 12 = 12$

2. $2x + 4x = 27 - 3x$

$4x$

$3x \qquad \boxed{+3x}$ inverse of $-3x$

$9x \qquad = 27$

$9\overline{\smash{)}27} = 3 \qquad x = 3$

Proof: $2(3) + 4(3) = 27 - 3(3)$
$\qquad\quad 6 \;+\; 12 \;= 27 - \;\; 9$
$\qquad\qquad 18 \quad = \quad 18$

REVIEW QUESTIONS

6.25. Solve the following equations.

(a) $x^2 + 4 = 20$

(b) $16 = x + 3 + 2 - 1$

(c) $2x - 3 = 22$

(d) $4x = 40$

(e) $2x + 3x = 36 - 4x$

(f) $\frac{x}{3} = 8$

(g) $46 = \frac{x}{2}$

Distributive Law

In Chapter 4, the *distributive law* was defined as follows: If two numbers are to be added, then multiplied by a third number [$a(b + c)$], the result will be the same if each of the numbers to be added is multiplied by the third number, then the individual products ($ab + ac$) added. In equation form, the distributive law may be written as follows:

$$a(b + c) = ab + ac$$

The law holds true regardless of the number of elements:

$$a(b + c + d + e) = ab + ac + ad + ae$$

This law can be used to simplify addition in equations. For example, if $3x$ and $4x$ are on the same side of an equation, it would be simpler to write $7x$ and it would mean exactly the same thing. On the other hand, if $3x$ and $4y$ were on the same side of the equation, no simplification could be accomplished.

EXAMPLES

$$2x + 4x + 3y \qquad \text{simplify to: } 6x + 3y$$
$$3x + 4y - 1x \qquad \text{simplify to: } 2x + 4y$$
$$2xy + 3xy \qquad \text{simplify to: } 5xy$$
$$\tfrac{1}{3}x + \tfrac{1}{4}x + 2x \qquad \text{simplify to: } 2\tfrac{7}{12}x$$

When the symbols are alike, as in the above examples, they are called *like terms* and may be added (or subtracted) together, thus simplifying equations.

REVIEW QUESTIONS

6.26. Simplify the following expressions, if possible.

(a) $2x - x + 4x$ (c) $2x + 2y + 2z$

(b) $3x + 3y + 4x$ (d) $4x - 2y - 2x$

6.27. Using the numbers 2 for x, 3 for y, and 4 for z, solve the expressions in problem 6.26 above.

6.28. If five times a number plus four times the same number plus three times the same number equals 72, what is the number?

6.29. Solve the following equations for the unknown element.

(a) $3.6x + 4.2x - 2x + x = 68$ (c) $2x + 4 + 2x = 48$

(b) $3x + 2x + 4 - 2 = 60$ (d) $4x - 2x + 3x - 3 = 42$

6.30. What is the root of the following equations?

(a) $\frac{1}{4}x + 2x + 1.8x = 40$

(b) $6x + 5 - 2x + 3 = 46$

CONDITIONAL EQUATIONS

In our discussion of equations, we have been mainly interested in establishing the truth of an equation and in solving for unknowns in an equation.

The expression $3x + 6$ does not, in itself, imply that x must be any particular number. Only if the equation is completed will specific limits be given to x.

EXAMPLES

1. $3x + 6 = 24$

Now x must be 6 to satisfy the requirements of the total equation.

2. $3x + 6 < 14$

Now x will be any number in the set $\{0, 1, 2\}$ to satisfy the requirements of the statement.

Conditional equations are those which require one or more unknowns to be solved. These types of equations can be very useful in solving real life problems, but the difficult part of the process is the transformation of word-oriented problems into equations which can easily be solved by the computer. One such word problem is problem 6.28 in the previous review section. This is an extremely simple problem, and under normal circumstances such a problem would not be programmed for computer processing.

On the other hand, if a problem can be solved by simple algebraic equation manipulation, the work of setting up the problem falls on the shoulders of the programmer and he must know how to solve the problem before he can program it for the computer to process. The computer is particularly good at doing simple, repetitive jobs at an extremely fast pace. It is this facility that has made it such a necessary tool for the business man, the scientist, and the mathematician.

Telling the computer what to do and how to do it is the function of the computer programmer. He must know not only the language of the computer so that he can communicate with it, but he must also be able to do problem analysis so that the best methods of attacking a problem will be used. It is in this area that a knowledge of functions and equations are invaluable.

EXAMPLE

Input is a deck of punched cards, each card containing an amount for variables a and b. For each pair of variables, calculate the value of c, using the following equation.

$$c = \sqrt{a^2 + b^2}$$

Output will be a report of all c values.

A flow chart of the steps to accomplish the process could be as follows:

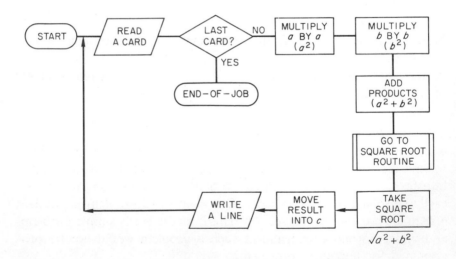

As soon as the flow chart has been developed, it is a small step to writing the program that will accomplish the process. In this example, there was no unknown to find, but the method of solution would be the same in either case.

Assume that the output report shows a series of "c" measurements (13, 11, 14, 8, 4, 13, 14, 12) and that this covers a time frame of one working day, with one measurement being taken every hour and the expected measurement being $12 \pm$ (plus or minus) 2. This could be shown in graph form so that the data would be more easily understandable.

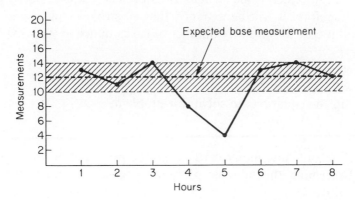

Notice how the measurements dropped off between the fourth and fifth hour, then recovered between the fifth and sixth hour. It is this type of timely information that allows a manager to make decisions based on knowledge of situations instead of using intuition and guesswork.

REVIEW QUESTIONS

6.31. Solve the following problem with a step-by-step flow chart. Input is a card deck; each card contains fields A, B, and C. The program is to compute the following formula:

$$X = A + (B * C)$$

Output will be a deck of cards containing the quantity Y in each card. Y is derived from the equation

$$Y = 20 - 2X$$

In the previous example and problem, the equations were specified, simplifying the job of analysis. Most problems are not as clean-cut and the equations must be developed by the programmer, based upon the information available to him.

EXAMPLE

The cost of keypunching cards is $4.00 an hour. There are two operators available; operator 1 can punch 8,000 strokes an hour and operator 2 can punch 6,400 strokes an hour. There are 80 positions on a card; therefore, it will be assumed that 80 strokes equals one card. What will it cost to have any number of cards punched (use 500 as an example):

1. with operator 1?
2. with operator 2?

Set up the equations to solve this problem.

Analysis:

1. Find out how many cards operator produces in one hour.
2. Determine the number of hours needed to punch the total number of cards.
3. Multiply the number of hours by $4.00 to get total cost.

Symbols:

$$a = \text{number of strokes to one card}$$
$$b = \text{number of strokes per hour}$$
$$c = \text{number of cards produced per hour}$$
$$d = \text{total number of cards needed}$$
$$e = \text{number of hours punching required}$$
$$t = \text{total cost}$$

Equations:

$$\frac{b}{a} = c \qquad \frac{d}{c} = e \qquad e(4.00) = t$$

Solution for operator 1: $\dfrac{8,000}{80} = 100$ $\dfrac{500}{100} = 5$ $5(4.00) = \$20.00$

Solution for operator 2: $\dfrac{6,400}{80} = 80$ $\dfrac{500}{80} = 6.25$ $6.25(4.00) = \$25.00$

There could be any number of variations on this problem using both operators in various ways to get the optimum work out of the entire keypunching team. Some of these ideas will be developed later in the chapter.

The *analysis* portion of the job specified *what* must be accomplished and the *order* in which steps are to be accomplished. Symbols are then set up to represent *known* elements and elements that must be *determined* in the logical sequence of events. The equations to solve the various portions of the problem are set up and sample solutions are computed to prove the equations. When all of this has been accomplished, the final steps of flow charting and writing the program are quite simple and straightforward.

REVIEW QUESTIONS

6.32. In the previous example, which items were *variable*? What was the *constant*?

6.33. Cost of using the computer is $100.00 per hour. It takes 1 minute, 10 seconds to process one document. Add 1 minute, 30 seconds to each total job for setup and takedown time. How much will it cost to process a job?

LINEAR EQUATIONS

We have mentioned earlier in the chapter that a function is called *linear* if a graph of the function forms a straight line. Linear equations have the same properties. There must be at least two points that satisfy the equation for a line to be plotted on a graph.

EXAMPLE

$$x + 2y = 6$$

x could be 0 and y could be 3 to satisfy the equation; x could be 2 and y could be 2 to satisfy the equation. We now have two points: $(0, 3)$ $(2, 2)$.

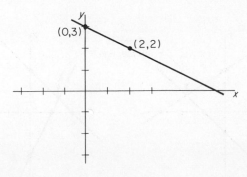

Simultaneous Equations

Often, it is useful to examine two equations at the same time to look for solutions that are common to both.

EXAMPLES

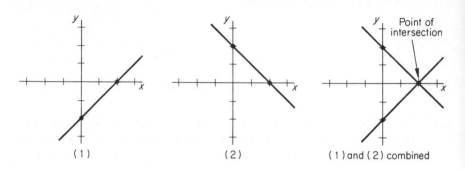

Examination of two linear equations in this manner is called the *simultaneous system* of linear equations. Solution of these types of equations may be accomplished graphically or algebraically. The graphical method is much faster, but it is possible to make errors in graphing and occasionally it may be difficult to get an exact value for a point of intersection.

Graphical Solution

The graphical method is based on the assumption that when two straight lines intersect, there can be one, and *only* one, point of intersection. That point is common to the two equations.

EXAMPLES

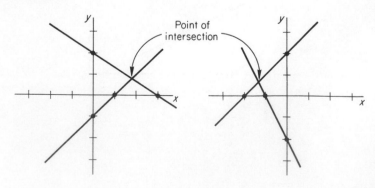

You will remember that when coordinates are specified in pairs, the first number represents the domain (x axis) and the second number represents the range (y axis).

EXAMPLE

$$(-2, -2)\ (0, 0)\ (2, 2)\ (3, 3)$$

These represent points on the graph:

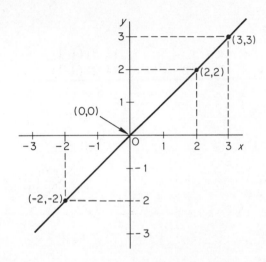

As in the example above, two linear equations can be plotted on the same graph, but there must be at least two points for each equation.

EXAMPLE

Equation 1:	$x + 2y = 8$	$(2, 3)\ (4, 2)$
Equation 2:	$2x - y = 6$	$(2, -2)\ (1, -4)$

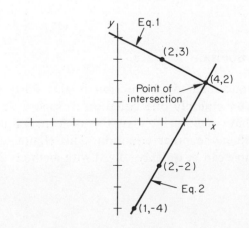

The problem is worked by finding the two points which satisfy equation 1 and the two points which satisfy equation 2. If the two lines do not intersect, there is no common point in the two equations. This can only happen if the two lines are exactly parallel.

The point of intersection in the example above is the solution to the problem and represents the *only* values of x and y which satisfy both equations. In this example the solution is (4, 2).

EXAMPLE

| *Equation 1:* | $x + y = 5$ | (2, 3) (4, 1) |
| *Equation 2:* | $2x - 2y = 4$ | (4, 2) (3, 1) |

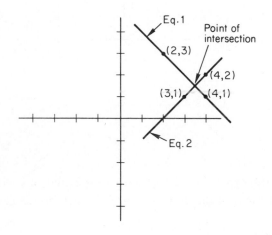

The solution is (3.5, 1.5). Some solutions are hard to estimate with a graph. For such problems, another method of solution is more practical.

Algebraic Solution

The algebraic method of solution is also fairly simple. Find a constant that can multiply one equation in such a way that the new equation will have one variable of the same magnitude, but with a different sign than the other equation. This eliminates one variable. The other variable can be quickly solved with methods described on the previous pages.

EXAMPLE 1

$$\begin{array}{ll} \textit{Equation 1:} & x + 2y = 8 \\ \textit{Equation 2:} & 2x - y = 6 \end{array} \longrightarrow \text{multiply by 2:}$$

$$\begin{array}{l} x + 2y = 8 \\ 4x - 2y = 12 \\ \hline \end{array}$$

$$\text{then add} \longrightarrow \quad 5x + 0 = 20$$

$$x = 4 \qquad\qquad\qquad 5(4) + 0 = 20$$

Then back to either of the original equations to solve for y:

1. $x + 2y = 8$

   $4 + 2(2) = 8 \qquad y = 2$

2. $2x - y = 6$

   $2(4) - 2 = 6 \qquad y = 2$

EXAMPLE 2

$$\textit{Equation 1:} \qquad \begin{array}{l} x + y = 5 \quad (x3) \longrightarrow 3x + 3y = 15 \\ 2x - 3y = 2 \qquad\qquad\quad \underline{2x - 0 = 2} \\ \qquad\qquad\qquad\qquad\qquad 5x + 0 = 17 \\ \qquad\qquad\qquad\qquad\qquad\quad x = 3.4 \end{array}$$

$$3.4 + y = 5$$
$$y = 1.6$$

Proof:

$$\textit{Equation 2:} \quad \begin{array}{l} 2(3.4) - 3(1.6) = 2 \\ 6.8 - 4.8 = 2 \end{array}$$

A practical application for the solution of simultaneous equations could be the keypunching example developed on page 148, with one additional piece of information added.

It is known that operator 2 can punch 80 cards per hour and operator 1 can punch 100 cards per hour. Also, the charge for keypunching is $4.00 per hour. Assume a job is received which requires the punching of 500 cards for $22.00. It is necessary to determine how many cards each operator must punch in order to accomplish the task for the $22.00 commitment. One equation that can be derived from the above statement is:

Equation 1: $x + y = 500$

where

$x =$ number of cards punched by operator 2

$y =$ number of cards punched by operator 1

500 = total number of cards to be punched

Another equation that can be derived is:

Equation 2: $\dfrac{x}{80}(4.00) + \dfrac{y}{100}(4.00) = 22.00$

or

$$.05x \quad + \quad .04y \quad = 22.00$$

where

$x =$ number of cards punched by operator 2
$y =$ number of cards punched by operator 1
$80 =$ number of cards operator 2 can punch per hour
$100 =$ number of cards operator 1 can punch per hour
$4.00 =$ dollars per hour charge for keypunching
$22.00 =$ amount charged for keypunching job

Solving equations (1) and (2) simultaneously,

$$x + \quad y = 500 \tag{1}$$
$$.05x + .04y = 2.200 \tag{2}$$

Multiplying equation (1) by $-.04$,

$$-.04x - .04y = 20.00 \tag{1}$$
$$\underline{.05x + .04y = 22.00} \tag{2}$$
$$.01x \qquad = 2.00$$
$$x \qquad = 200$$

Substituting $x = 200$ in equation (1),

$$200 + y = 500$$
$$y = 300$$

Therefore, operator 2 must punch 200 cards and operator 1 must punch 300 cards in order to meet the cost commitment of $22.00 for 500 cards.

REVIEW QUESTIONS

6.34. Solve for the point of intersection for the following two equations, using the graphical method.
Equation 1: $2x + 4y = 8$ (0, 2) (4, 0)
Equation 2: $3x - 2y = 12$ (0, −6) (4, 0)

6.35. Solve the equations in problem 6.34, using the algebraic method.

More Than Two Linear Equations

Either the graphic or the algebraic solution is feasible for more than two linear equations. It looks a little more messy, but it can still be handled on a two-by-two basis.

EXAMPLE

Equation 1: $x + 2y = 8$ (4, 2) (2, 3)
Equation 2: $2x - y = 5$ (4, 3) (3, 1)
Equation 3: $3x - 2y = 5$ (3, 2) (2, .5)

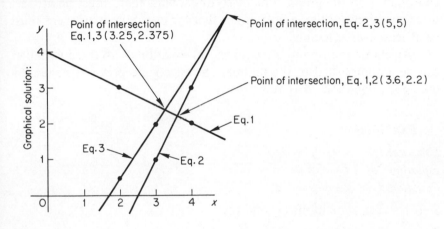

Algebraic solution:

Equation 1: $x + 2y = 8$
Equation 2: $2x - y = 5$

$$
\begin{array}{l}
x + 2y = 8 \\
(\times 2) \longrightarrow 4x - 2y = 10 \\
\hline
5x + 0 = 18 \qquad x = 3.6 \\
3.6 + 2y = 8 \qquad y - 2.2
\end{array}
$$

point of intersection, equations (1), (2), (3.6, 2.2)

Equation 1: $x + 2y = 8$ multiplication is not necessary since one
Equation 3: $3x - 2y = 5$ element can automatically be eliminated

$$
\begin{array}{l}
\hline
4x + 0 = 13 \qquad x = 3.25 \\
3.25 + 2y = 8 \qquad y = 2.375
\end{array}
$$

point of intersection, equations (1), (3), (3.25, 2.375)

Equation 2: $2x - y = 5$ $x(-2) \longrightarrow -4x + 2y = -10$
Equation 3: $3x - 2y = 5$ $3x - 2y = 5$

$$5x - 3y = 10$$ $$\begin{aligned} -x + 0 &= -5 & x &= 5 \\ 2(5) - y &= 5 & y &= 5 \end{aligned}$$

point of intersection, equations (2), (3), (5, 5)

Equations With Three or More Variables

The two-dimensional graphical method of solution is not possible if the equation has more than two variables, since there are only two coordinates on the graph. For three unknowns, there must be at least three equations, in the same manner that for two unknowns there must be at least two equations.

Algebraic solution to three (or more) equations with three variables can be accomplished in the manner shown on the previous pages, by taking the equations two at a time.

EXAMPLE

Equation 1: $x + y + 2z = 8$
Equation 2: $2x - 3y + z = 1$
Equation 3: $x + 2y - z = -7$

Step 1: Take equations (1) and (2):

Equation 1: $x(3)$ $3x + 3y + 6z = 24$
Equation 2: $2x - 3y + z = 1$

Result 1: $\overline{5x + 0 + 7z = 25}$

Step 2: Take equations (1) and (3):

Equation 1: $x(-2)$ $-2x - 2y - 4z = -16$
Equation 3: $x + 2y - z = -7$

Result 2: $\overline{-x + 0 - 5z = -23}$

Step 3: Solve equations result 1 and result 2:

$$\begin{aligned} 5x + 7z &= 25 \\ x(5)\quad -5x - 25z &= -115 \\ \hline 0 - 18z &= -90 \\ z &= 5 \end{aligned}$$

(We now have solved one of the three variables.)

Step 4:   Substituting in result 2,
$$-x - 5(5) = -23$$
$$x = -2$$

Step 5:   Substituting in equation (1),
$$-2 + y + 10 = 8$$
$$y = 0$$

Solution: $x = -2, y = 0, z = 5$

Proof:
Equation 1: $-2 + 0 + 10 = 8$
Equation 2: $-4 - 0 + 5 = 1$
$$1 = 1$$
Equation 3: $-2 + 0 - 5 = -7$
$$-7 = -7$$

A little practice with this method will make the solution of these types of problems quite easy to accomplish.

REVIEW QUESTIONS

6.36. Find the solution(s) to the following system of linear equations.
Equation 1: $x - y + z = 3$
Equation 2: $3x + 2y - z = 0$
Equation 3: $2x + y + 2z = 3$

$$x =$$
$$y =$$
$$z =$$

REVIEW

Review of Terminology

Function	the relationship between two sets where each element of the first set is related to just one element of the second set
Domain	the first set in a function
Range	elements of the second set that are assigned to elements of the first set

Mapping	showing the relationship of elements of the domain and range of the function
Image	the element of the range as it relates to the element of the domain
Discontinuous	specific points on a graph, where no additional measurements are possible
Continuous	a smooth line on a graph upon which continuous measurements may be taken
Point of origin	the zero point on a graph
Quadrants	the four rectangles formed by a Cartesian coordinate graph
Sentences	statements written in symbols
Open sentence	algebraic statement containing one or more missing element(s)
Replacement set	a set of numbers, each of which can replace an unknown in an equation
Variable	an element in an equation that will have more than one value associated with it
Root	a number that replaces a symbol which was an unknown in an equation
Additive inverse	adding the inverse of a known element to the other side of the equation, to simplify the equation (this is really adding to both sides)
Multiplicative inverse	multiplying both sides of an equation with the inverse of a known element, to simplify the equation
Like terms	two or more symbols that are alike
Constant	an element used in an equation that does not change in value

7 MATRICES

DESCRIPTION AND DEFINITIONS

A matrix is an array of numbers (or symbols) set up in a sequence of rows and columns. If letters are used, they represent real numbers. Just as in sets, matrices are given names with capital letters.

In a matrix, *rows* are horizontal and *columns* are vertical. The number of rows or columns in a matrix is called the *size* of the matrix.

A matrix is enclosed in brackets.

EXAMPLE

$$A = \begin{bmatrix} 1 & 3 & 4 & 5 \\ 6 & 8 & 9 & 11 \end{bmatrix}$$

The size of matrix A is two rows and four columns. This can be expressed as a 2×4 matrix.

$$B = \begin{bmatrix} a & b & c \\ d & e & f \end{bmatrix}$$

The size of matrix B is two rows and three columns. This can be expressed as a 2 × 3 matrix.

When letters are used for elements of a matrix, it is usual to indicate the position of each element with two numbered subscripts; the first one for row and the second one for column.

EXAMPLE

$$A = \begin{bmatrix} a_{11} & a_{12} & a_{13} \\ a_{21} & a_{22} & a_{23} \\ a_{31} & a_{32} & a_{33} \end{bmatrix}$$

Notation for any matrix is usually limited to the small alphabetic character of the name of the matrix, with appropriate subscripts.

When a matrix has an identical number of rows and columns, it is called a *square matrix*. If a matrix consists of just one row, it is called a *row matrix* and if it contains just one column, it is called a *column matrix* (these last two are also called *vectors*).

EXAMPLES

square matrix

$$A = \begin{bmatrix} 1 & 2 & 3 & 4 \\ 3 & 6 & 9 & 8 \\ 10 & 12 & 11 & 7 \\ 4 & 8 & 2 & 3 \end{bmatrix}$$

row matrix
(row vector)

$$B = [b_1\ b_2\ b_3\ b_4]$$

column matrix
(column vector)

$$C = \begin{bmatrix} c_1 \\ c_2 \\ c_3 \\ c_4 \end{bmatrix}$$

A square matrix has an imaginary line from the top left corner to the bottom right corner, called the *main diagonal* of the matrix.

EXAMPLE

$$A = \begin{bmatrix} a_{11} & a_{12} & a_{13} \\ a_{21} & a_{22} & a_{23} \\ a_{31} & a_{32} & a_{33} \end{bmatrix}$$

Notice that the main diagonal is at the point in each row and column where row equals column (row 1, column 1; row 2, column 2; row 3, column 3).

It must be understood that matrices do not have values in themselves. Matrices are used as a means of putting numbers into table form. The matrix helps to describe a system of equations, which then can be added, subtracted, or multiplied in particular ways.

EXAMPLE

Equation 1:

$$x + y - 2z = 8$$

Equation 2:

$$2x - 3y + z = 1$$

Equation 3:

$$x + 2y - z = -7$$

In matrix form:

$$A = \begin{bmatrix} 1 & 1 & -2 \\ 2 & -3 & 1 \\ 1 & 2 & -1 \end{bmatrix}$$

This particular matrix is meaningless, except as it relates to the variables x, y, and z and the right-hand side of each equation $(8, 1, -7)$.

Solving simultaneous systems of equations (discussed in Chapter 6) is simple enough as long as only a few equations are involved, but if a great number of equations are to be solved, the use of matrices and computer processing become necessary. This is the type of operation that really allows the computer to shine since each individual operation is simple, but the volume of such simple operations makes it awkward (or even impossible) to do by hand.

REVIEW QUESTIONS

7.1. What is the size of the following matrices?

(a) $A = \begin{bmatrix} a & b & c \\ d & e & f \end{bmatrix}$

(b) $B = \begin{bmatrix} 1 & 2 & 3 & 4 \end{bmatrix}$

(c) $C = \begin{bmatrix} 1 & 2 \\ 3 & 4 \end{bmatrix}$

7.2. (a) Which of the matrices in problem 7.1 can be called a square matrix?
(b) Is there a row or column matrix in problem 7.1? Which one?
(c) What is the main diagonal of the square matrix in problem 7.1?

7.3. Construct a square matrix of the elements:

$$a, b, c, d, e, f, g, h, i$$

Call it matrix X. Take the elements sequentially, by rows, and insert proper subscripts.

7.4. In matrix X (problem 7.3), which element has the subscripts?
(a) 21   (c) 32
(b) 23   (d) 13

7.5. In matrix X (problem 7.3), what are the names of the elements on the main diagonal?

Equal Matrices

It is possible to add, subtract, multiply, and invert matrices, but first an understanding of the various types of matrices must be developed.

Two matrices are *equal* if *all* corresponding elements in *both* matrices are equal.

EXAMPLES

1.
$$A = \begin{bmatrix} 1 & 2 & 3 \\ 4 & 5 & 6 \end{bmatrix} \qquad B = \begin{bmatrix} 1 & 2 & 3 \\ 4 & 5 & 6 \end{bmatrix}$$

Now we can say: $A = B$.

2.
$$A = \begin{bmatrix} 1 & 2 & 3 \\ 4 & 5 & 6 \end{bmatrix} \qquad B = \begin{bmatrix} b_{11} & b_{12} & b_{13} \\ b_{21} & b_{22} & b_{23} \end{bmatrix}$$

If it is specified that these two matrices are equal ($A = B$), this implies that:

$$b_{11} = 1, \qquad b_{12} = 2, \qquad b_{13} = 3$$
$$b_{21} = 4, \qquad b_{22} = 5, \qquad b_{23} = 6$$

3.
$$\begin{bmatrix} a_1 \\ a_2 \\ a_3 \end{bmatrix} = \begin{bmatrix} 4 + 2 - 1 \\ 3 - 1 + c \\ 2 + 3 - 1 \end{bmatrix}$$

This implies that:

$$a_1 = 5$$
$$a_2 = 2 + c$$
$$a_3 = 4$$

MATRIX ADDITION AND SUBTRACTION

The addition of matrices is dependent on equality, but not on equality of elements. There must be an equal number of rows and columns to perform matrix addition.

If two matrices have an equal number of rows and an equal number of columns, they are called *equivalent* and addition is accomplished by adding corresponding elements (element with subscript 11 to element with subscript 11, 12 to 12, 13 to 13, etc.).

EXAMPLE

$$A = \begin{bmatrix} a_{11} & a_{12} & a_{13} \\ a_{21} & a_{22} & a_{23} \end{bmatrix}$$

$$B = \begin{bmatrix} b_{11} & b_{12} & b_{13} \\ b_{21} & b_{22} & b_{23} \end{bmatrix}$$

$$\begin{aligned} A + B = a_{11} + b_{11} \\ a_{12} + b_{12} \\ a_{13} + b_{13} \\ a_{21} + b_{21} \\ a_{22} + b_{22} \\ a_{23} + b_{23} \end{aligned}$$

This forms a new matrix, which we can call C:

$$C = A + B = \begin{bmatrix} a_{11} + b_{11} & a_{12} + b_{12} & a_{13} + b_{13} \\ a_{21} + b_{21} & a_{22} + b_{22} & a_{23} + b_{23} \end{bmatrix}$$

It is easy to see that if matrices are of different sizes, it would not be possible to add corresponding elements throughout. Some elements in the larger matrix would be left over. An example of the method of addition using real numbers makes it very easy to follow the process.

EXAMPLE

$$A = \begin{bmatrix} 2 & -3 & 4 \\ 6 & 1 & -2 \end{bmatrix} \quad B = \begin{bmatrix} 3 & 2 & -1 \\ -2 & 3 & 4 \end{bmatrix}$$

$$\begin{aligned} 2 + 3 &= 5 & 6 + (-2) &= 4 \\ -3 + 2 &= -1 & 1 + 3 &= 4 \\ 4 + (-1) &= 3 & -2 + 4 &= 2 \end{aligned}$$

This results in:

$$C = A + B = \begin{bmatrix} 5 & -1 & 3 \\ 4 & 4 & 2 \end{bmatrix}$$

Subtraction of matrices follows the same procedure and has the same requirements imposed upon it. The rows and columns of the two matrices must be the same size; then an element of the second matrix is subtracted from the corresponding element of the first matrix, continuing until all elements have been subtracted. The result of the subtractions forms a new matrix.

EXAMPLE

Matrices:

$$A = \begin{bmatrix} 2 & -3 & 4 \\ 6 & 1 & -2 \end{bmatrix} \qquad B = \begin{bmatrix} 3 & 2 & -1 \\ -2 & 3 & 4 \end{bmatrix}$$

Computation:

$$
\begin{array}{llll}
A - B = & 2 - & 3 & = -1 & 6 - (-2) = & 8 \\
& -3 - & 2 & = -5 & 1 - & 3 & = -2 \\
& 4 - (-1) = & 5 & -2 - & 4 & = -6
\end{array}
$$

Result:

$$C = A - B = \begin{bmatrix} -1 & -5 & 5 \\ 8 & -2 & -6 \end{bmatrix}$$

REVIEW QUESTIONS

7.6. Which of the following matrices are equivalent and which are truly equal?

$$A = \begin{bmatrix} 1 & 2 & 3 \\ 4 & 5 & 6 \end{bmatrix} \qquad B = \begin{bmatrix} b_{11} & b_{12} & b_{13} \\ b_{21} & b_{22} & b_{23} \end{bmatrix}$$

$$C = \begin{bmatrix} c_{11} & c_{12} & c_{13} \\ c_{21} & c_{22} & c_{23} \\ c_{31} & c_{32} & c_{33} \end{bmatrix} \qquad D = \begin{bmatrix} d_{11} & d_{12} & d_{13} \\ d_{21} & d_{22} & d_{23} \end{bmatrix}$$

$$E = \begin{bmatrix} 1 & 2 & 3 \\ 4 & 5 & 6 \end{bmatrix} \qquad F = \begin{bmatrix} f_{11} & f_{12} & f_{13} \\ f_{21} & f_{22} & f_{23} \end{bmatrix}$$

7.7. If we specify that matrix D is equal to matrix A (in problem 7.6), what will be the elements of matrix D?

7.8. In problem 7.6, add matrix A and matrix E to form a new matrix G.

7.9. (a) Which matrix in problem 7.6 is a square matrix?
(b) What is its main diagonal?

7.10. What is the requirement for matrices to be added or subtracted?

7.11. Which of the matrices in problem 7.6 can be added together?

7.12. Add the following pairs of matrices. Show the resulting matrix from each pair being added.

(a) $A = \begin{bmatrix} 1 & 3 & 5 \\ 2 & 4 & 6 \end{bmatrix}$ $\qquad B = \begin{bmatrix} 2 & 3 & 4 \\ 6 & 1 & 2 \end{bmatrix}$ $\qquad C = A + B = ?$

(b) $P = \begin{bmatrix} p_{11} & p_{12} & p_{13} \\ p_{21} & p_{22} & p_{23} \end{bmatrix}$ $Q = \begin{bmatrix} q_{11} & q_{12} & q_{13} \\ q_{21} & q_{22} & q_{23} \end{bmatrix}$ $R = P + Q = ?$

(c) $U = \begin{bmatrix} 2 & -2 & -1 \\ 3 & -3 & 3 \end{bmatrix}$ $V = \begin{bmatrix} -1 & -2 & -3 \\ 3 & 2 & -2 \end{bmatrix}$ $W = U + V = ?$

7.13. Subtract the following pairs of matrices. Show the resulting matrix from each pair being subtracted.

(a) $A = \begin{bmatrix} 2 & -2 & -1 \\ 3 & -3 & 3 \end{bmatrix}$ $B = \begin{bmatrix} -1 & -2 & -3 \\ 3 & 2 & -2 \end{bmatrix}$ $C = A - B = ?$

(b) $P = \begin{bmatrix} 4 & -3 & x \\ -2 & 3 & -1 \end{bmatrix}$ $Q = \begin{bmatrix} -2 & -2 & -2 \\ 4 & -2 & -3 \end{bmatrix}$ $R = P - Q = ?$

MATRIX MULTIPLICATION

If a matrix is to be multiplied by a single number, each element in the matrix will be multiplied by that number. This is true no matter what the size of the matrix happens to be.

EXAMPLE

$$A = \begin{bmatrix} 2 & -3 & 2 \\ 4 & 1 & -2 \end{bmatrix}$$

Multiply matrix A by 3:

$$3A = \begin{bmatrix} 6 & -9 & 6 \\ 12 & 3 & -6 \end{bmatrix}$$

Suppose we had the following problem:

$$2A + B = C \quad \text{(where } A, B, \text{ and } C \text{ are matrices)}$$

$$A = \begin{bmatrix} 2 & -3 & 2 \\ 4 & 1 & -2 \end{bmatrix} \quad B = \begin{bmatrix} 3 & 2 & -2 \\ 2 & -2 & -3 \end{bmatrix}$$

The first step is to multiply matrix A by 2, then the resultant matrix is added to matrix B.

$$2A = \begin{bmatrix} 4 & -6 & 4 \\ 8 & 2 & -4 \end{bmatrix} \quad B = \begin{bmatrix} 3 & 2 & -2 \\ 2 & -2 & -3 \end{bmatrix}$$

$$C = 2A + B = \begin{bmatrix} 7 & -4 & 2 \\ 10 & 0 & -7 \end{bmatrix}$$

Two matrices can be multiplied if certain conditions are met. In this type of multiplication, elements are multiplied by rows and columns; therefore, the first matrix must have as many columns as the second matrix has rows. In the example above, matrix A has *three columns* and matrix B has *two rows;* therefore, they cannot be multiplied together.

EXAMPLE

$$A = \begin{bmatrix} 2 & -3 & 2 \\ 4 & 1 & -2 \end{bmatrix} \quad B = \begin{bmatrix} 3 & 3 & -2 \\ 2 & -2 & -3 \\ 4 & 2 & -2 \end{bmatrix}$$

Matrix A has *three columns*. Matrix B has *three rows*. The conditions are met and the two matrices can be multiplied. There is no limit to the number of *rows* in matrix A and no limit to the number of *columns* in matrix B.

EXAMPLE

$$A = \begin{bmatrix} 2 & -3 & 2 \\ 4 & 1 & -2 \\ -3 & 2 & -1 \\ 5 & -1 & 4 \end{bmatrix} \quad B = \begin{bmatrix} 3 & 2 & -2 & 4 & 6 \\ 2 & -2 & -3 & -1 & 3 \\ 4 & 2 & -2 & 3 & -2 \end{bmatrix}$$

A is a 4×3 matrix; B is a 3×5 matrix. Note the two numbers that must be identical: $4 \times \underline{3}; \underline{3} \times 5$

The actual multiplication process is really a combination of multiplication and addition, resulting in a new matrix containing the multiplied numbers and with a size that equals the number of *rows* in *A* and the number of *columns* in *B*. The new matrix formed by the example above will be 4×5.

REVIEW QUESTIONS

7.14. Can a pair of square matrices be multiplied together?

7.15. Which of the following pairs of matrices can be multiplied?
(a) 4×3 and 3×4 (d) 4×2 and 4×3
(b) 5×4 and 6×4 (e) 6×3 and 3×6
(c) 4×4 and 4×4

7.16. What will be the size of the resulting matrix for each pair that can be multiplied in problem 7.15?

7.17. Solve the following:
(a) $A = \begin{bmatrix} -4 & 4 & -2 \\ 2 & -3 & -1 \end{bmatrix}$ $4A = ?$

(b) $A + 3B = C$

$A = \begin{bmatrix} 3 & -2 & 4 \\ -2 & 1 & -3 \end{bmatrix}$ $B = \begin{bmatrix} -2 & 3 & -1 \\ 4 & -2 & -3 \end{bmatrix}$

Method of Multiplying Matrices

Multiplying matrices is a time consuming and laborious job. It is the type of work that is ideally suited to the capabilities of a computer. The simplest way to show the method is with an example.

EXAMPLE

$$A = \begin{bmatrix} a_{11} & a_{12} \\ a_{21} & a_{22} \end{bmatrix} \qquad B = \begin{bmatrix} b_{11} & b_{12} \\ b_{21} & b_{22} \end{bmatrix}$$

The elements of each *row* of matrix *A* are multiplied by the elements of each *column* of matrix *B*, summing the results of each row-column multiplication.

$$a_{11} \times b_{11} + a_{12} \times b_{21} = \text{new element}_{11}$$
$$a_{11} \times b_{12} + a_{12} \times b_{22} = \text{new element}_{12}$$
$$a_{21} \times b_{11} + a_{22} \times b_{21} = \text{new element}_{21}$$
$$a_{21} \times b_{12} + a_{22} \times b_{22} = \text{new element}_{22}$$

It is easier to see the process when numbers are used:

$$A = \begin{bmatrix} 3 & 4 \\ 2 & 3 \end{bmatrix} \qquad B = \begin{bmatrix} 5 & 6 \\ 7 & 8 \end{bmatrix}$$

Step 1:

$$A = \begin{bmatrix} 3 & 4 \\ 2 & 3 \end{bmatrix} \qquad B = \begin{bmatrix} 5 & 6 \\ 7 & 8 \end{bmatrix}$$

$$3 \times 5 + 4 \times 7 = 43$$

Step 2:

$$A = \begin{bmatrix} 3 & 4 \\ 2 & 3 \end{bmatrix} \qquad B = \begin{bmatrix} 5 & 6 \\ 7 & 8 \end{bmatrix}$$

$$3 \times 6 + 4 \times 8 = 50$$

Step 3:

$$A = \begin{bmatrix} 3 & 4 \\ 2 & 3 \end{bmatrix} \qquad B = \begin{bmatrix} 5 & 6 \\ 7 & 8 \end{bmatrix}$$

$$2 \times 5 + 3 \times 7 = 31$$

Step 4:

$$A = \begin{bmatrix} 3 & 4 \\ 2 & 3 \end{bmatrix} \qquad B = \begin{bmatrix} 5 & 6 \\ 7 & 8 \end{bmatrix}$$

$$2 \times 6 + 3 \times 8 = 36$$

$$3 \times 5 + 4 \times 7 = 43$$
$$3 \times 6 + 4 \times 8 = 50$$
$$2 \times 5 + 3 \times 7 = 31$$
$$2 \times 6 + 3 \times 8 = 36$$

Result of multiplication:

$$C = AB = \begin{bmatrix} 43 & 50 \\ 31 & 36 \end{bmatrix}$$

Another example, using a slightly larger pair of matrices, will help to clarify the process.

EXAMPLE

$$A = \begin{bmatrix} 3 & -2 & 4 \\ -2 & 1 & -3 \end{bmatrix} \qquad B = \begin{bmatrix} -2 & 3 & -1 \\ 4 & -2 & -3 \\ 2 & 1 & -2 \end{bmatrix}$$

A is $2 \times \underline{3}$; B is $\underline{3} \times 3$; they can be multiplied.

$$c_{11} = 3 \times (-2) + (-2) \times 4 + 4 \times 2 = -6$$
$$c_{12} = 3 \times 3 + (-2) \times (-2) + 4 \times 1 = 17$$
$$c_{13} = 3 \times (-1) + (-2) \times (-3) + 4 \times (-2) = -5$$

This completes the first row of matrix A with all three columns of matrix B.

$$c_{21} = -2 \times (-2) + 1 \times 4 + (-3) \times 2 = 2$$
$$c_{22} = -2 \times 3 + 1 \times (-2) + (-3) \times 1 = -11$$
$$c_{23} = -2 \times (-1) + 1 \times (-3) + (-3) \times (-2) = 5$$

This completes the second row of matrix A with all three columns of matrix B. The result will be a new matrix C.

$$C = AB = \begin{bmatrix} -6 & 17 & -5 \\ 2 & -11 & 5 \end{bmatrix}$$

The resultant matrix (C) is 2×3 as expected.

Think of the magnitude of the job if matrix A was 20×3 and matrix B was 3×15. There would have to be three hundred equations, each with three multiplications and additions. The computer can accomplish the job in the flash of an eye.

It is not possible to find the product of AB if the column-row rule is not met. In some instances, it is possible to find the product of BA. Usually, it will not be the same as the product of AB.

EXAMPLE

$$A = \begin{bmatrix} 3 & -2 & -3 \\ -2 & 4 & -1 \end{bmatrix} \qquad B = \begin{bmatrix} -2 & 2 \\ 3 & -3 \end{bmatrix}$$

$(2 \times 3 \text{ matrix}) \qquad (2 \times 2 \text{ matrix})$

no match

Now reverse the matrices:

$$B = \begin{bmatrix} -2 & 2 \\ 3 & -3 \end{bmatrix} \qquad A = \begin{bmatrix} 3 & -2 & -3 \\ -2 & 4 & -1 \end{bmatrix}$$

$(2 \times 2) \qquad (2 \times 3)$

match

$$
\begin{array}{rcr}
-2 \times & 3 + & 2 \times & -2 = -10 \\
-2 \times (-2) + & 2 \times & 4 = & 12 \\
-2 \times (-3) + & 2 \times (-1) = & 4
\end{array}
$$

$$
\begin{array}{rcr}
3 \times & 3 + (-3) \times (-2) = & 15 \\
3 \times (-2) + (-3) \times & 4 = -18 \\
3 \times (-3) + (-3) \times (-1) = & -6
\end{array}
$$

$$
C = BA = \begin{bmatrix} -10, & 12, & 4 \\ 15, & -18, & -6 \end{bmatrix}
$$

Earlier in the chapter, we multiplied two simple 2×2 matrices:

$$
A = \begin{bmatrix} 3 & 4 \\ 2 & 3 \end{bmatrix} \qquad B = \begin{bmatrix} 5 & 6 \\ 7 & 8 \end{bmatrix} \qquad C = AB = \begin{bmatrix} 43 & 50 \\ 31 & 36 \end{bmatrix}
$$

Reverse the matrices to prove that $AB \neq BA$:

$$
B = \begin{bmatrix} 5 & 6 \\ 7 & 8 \end{bmatrix} \qquad A = \begin{bmatrix} 3 & 4 \\ 2 & 3 \end{bmatrix}
$$

$$
\begin{array}{l}
5 \times 3 + 6 \times 2 = 27 \\
5 \times 4 + 6 \times 3 = 38 \\
\hline
7 \times 3 + 8 \times 2 = 37 \\
7 \times 4 + 8 \times 3 = 52
\end{array}
\qquad
C = BA = \begin{bmatrix} 27 & 38 \\ 37 & 52 \end{bmatrix}
$$

Vector Multiplication

Multiplying with a row matrix (row vector) follows the same rules as other matrix multiplication, but if a row and a column matrix are multiplied, the result is not a matrix, but a single number.

EXAMPLES

1. $\qquad A = \begin{bmatrix} 2 & 3 & 4 \end{bmatrix}$

   $(1 \times 3 \text{ matrix})$ $\qquad B = \begin{bmatrix} 2 & 1 \\ 3 & 2 \\ 4 & 3 \end{bmatrix}$

   $(3 \times 2 \text{ matrix})$

$$
2 \times 2 + 3 \times 3 + 4 \times 4 = 29 \\
2 \times 1 + 3 \times 2 + 4 \times 3 = 20
$$

$$
C = AB = \begin{bmatrix} 29 & 20 \end{bmatrix}
$$

$(1 \times 2 \text{ matrix})$

2. $$A = [4 \quad 3 \quad 2]$$
    $$(1 \times 3 \text{ matrix})$$
    $$B = \begin{bmatrix} 2 \\ 3 \\ 4 \end{bmatrix}$$
    $$(3 \times 1 \text{ matrix})$$

The result should be a 1×1 matrix—which is a single number:

$$4 \times 2 + 3 \times 3 + 2 \times 4 = 25$$

In this case:

$$C = AB = 25$$

3. Notice what happens when the matrices are reversed:

$$A = \begin{bmatrix} 2 \\ 3 \\ 4 \end{bmatrix} \qquad\qquad B = [4 \quad 3 \quad 2]$$
$$(3 \times 1 \text{ matrix}) \qquad\qquad (1 \times 3 \text{ matrix})$$

The resulting matrix (C) will be 3×3:

$2 \times 4 = 8$	$3 \times 4 = 12$	$4 \times 4 = 16$
$2 \times 3 = 6$	$3 \times 3 = 9$	$4 \times 3 = 12$
$2 \times 2 = 4$	$3 \times 2 = 6$	$4 \times 2 = 8$
(first row)	(second row)	(third row)

$$C = AB = \begin{bmatrix} 8 & 6 & 4 \\ 12 & 9 & 6 \\ 16 & 12 & 8 \end{bmatrix}$$

REVIEW QUESTIONS

7.18. (a) What would be the size of the resulting matrix if the two matrices being mutiplied were 20×3 and 3×15?

(b) How many elements would there be in such a matrix?

7.19. Multiply the following matrices and show the resulting matrix.

(a) $A = \begin{bmatrix} -3 & 2 \\ 4 & -3 \end{bmatrix}$ $\qquad B = \begin{bmatrix} -2 & 3 & -1 & 4 \\ -2 & -3 & 2 & 1 \end{bmatrix}$

$C = AB = \, ?$

(b) $A = \begin{bmatrix} 3 & -2 & 4 \\ -2 & 1 & -3 \end{bmatrix}$ $\qquad B = \begin{bmatrix} -2 & 3 \\ 4 & 1 \\ 2 & -2 \end{bmatrix}$

$C = AB = \, ?$

(c) $A = \begin{bmatrix} 3 & -3 & 2 \\ 2 & -2 & -1 \end{bmatrix}$ $B = \begin{bmatrix} -3 & 4 \\ 2 & -2 \end{bmatrix}$

$C = AB = ?$

7.20. Find the products of the following pairs of matrices.

(a) $A = [-3 \quad 2 \quad 4]$ $B = \begin{bmatrix} -2 & -3 & 4 \\ 2 & -1 & 3 \\ -4 & 2 & 3 \end{bmatrix}$

$C = AB = ?$

(b) $A = [4 \quad -2 \quad 3]$ $B = \begin{bmatrix} -1 \\ -3 \\ -2 \end{bmatrix}$

$C = AB = ?$

7.21. Find the products of both $A \times B$ and $B \times A$ for the following pairs of matrices.

(a) $A = \begin{bmatrix} -2 & 3 \\ 4 & -1 \end{bmatrix}$ $B = \begin{bmatrix} 3 & -4 \\ -2 & 3 \end{bmatrix}$

$C = AB = ?$
$D = BA = ?$

(b) $A = \begin{bmatrix} 2 \\ -3 \\ 4 \end{bmatrix}$ $B = [-2 \quad -3 \quad -4]$

$C = AB = ?$
$D = BA = ?$

DETERMINANTS

A great deal of matrix manipulation must be learned for the purpose of solving simultaneous linear equations in matrix form. Also, matrices play an important part in linear programming (Chapter 8).

So far, we have learned to add, subtract, and multiply matrices. It is not possible to divide matrices; therefore, the basic arithmetic functions have been covered.

In a 2×2 square matrix, the result of subtracting the opposite diagonal from the main diagonal, after multiplying the elements of each of the diagonals, is called a *determinant*.

EXAMPLES

1. $A = \begin{bmatrix} 4 & 3 \\ 2 & 5 \end{bmatrix}$

main diagonal: $4 \times 5 = 20$
opposite diagonal: $3 \times 2 = 6$
determinant: $\overline{14}$

The symbol for determinant is a pair of vertical lines.

$|A| = 14$ (The determinant of matrix A equals 14.)

2. $$B = \begin{bmatrix} b_{11} & b_{12} \\ b_{21} & b_{22} \end{bmatrix}$$

For any pair of 2×2 matrices, the equation can be written:

$$|B| = (b_{11} \times b_{22}) - (b_{12} \times b_{21})$$

The abbreviation for determinant is the capital letter D. To avoid any confusion, no matrix should be named "D," then any time "D" is used, it refers to "determinant" and not to a matrix.

Inconsistency and Dependency

If the determinant of a matrix is zero, there is no solution for the variables in a linear equation system. If this occurs, the equations are called *inconsistent*. Inconsistency may be compared to parallel linear equations that have no point of intersection.

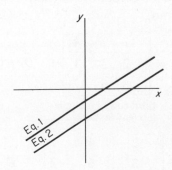

Equations are called *dependent* if they form the same line on a graph so that there are an infinite number of intersections.

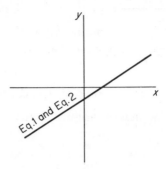

There is no solution for systems of linear equations that are *inconsistent* or *dependent*. An examination of any square matrix (regardless of size) will show that the determinant will be zero if any one of three conditions are met. Examples will be shown with 3×3 matrices.

Condition 1:   All three elements in a row or a column are zero.

$$A = \begin{bmatrix} 1 & 0 & 2 \\ 3 & 0 & -1 \\ -2 & 0 & -3 \end{bmatrix} \qquad D = 0$$

$$B = \begin{bmatrix} 0 & 0 & 0 \\ 2 & -1 & 3 \\ -1 & 2 & 1 \end{bmatrix} \qquad D = 0$$

Condition 2:   One row or column is an exact multiple of another row or column.

$$A = \begin{bmatrix} 1 & 2 & 4 \\ -2 & 3 & 1 \\ 3 & 6 & 12 \end{bmatrix} \qquad D = 0$$

third row is multiple
($3\times$) of first row

$$B = \begin{bmatrix} 2 & 4 & 1 \\ 3 & 6 & -2 \\ -4 & -8 & 3 \end{bmatrix} \qquad D = 0$$

second column is multiple
($2\times$) of first column

Condition 3: Two rows or columns are identical.

$$A = \begin{bmatrix} 1 & 2 & 1 \\ -2 & 3 & -2 \\ 4 & 2 & 4 \end{bmatrix} \quad D = 0$$

$$B = \begin{bmatrix} 2 & -1 & 3 \\ 2 & -1 & 3 \\ 1 & 3 & -2 \end{bmatrix} \quad D = 0$$

Solving Simultaneous Equations With Determinants

For systems of equations where solutions exist, determinants are often useful in reaching solutions. To solve for two variables (x and y), the following basic equations can be used:

$$x = \frac{Dx}{D} \quad y = \frac{Dy}{D}$$

The following example shows a way of solving simultaneous linear systems with the above equations, but first the determinants D, Dx, and Dy must be obtained. The equation setup for Dx and Dy should be examined very carefully in the following example.

EXAMPLE

Equation 1:

$$x + 2y = 8$$

Equation 2:

$$2x - y = 6$$

Step 1: Solve for D.

$$D = \begin{bmatrix} 1 & 2 \\ 2 & -1 \end{bmatrix} = 1(-1) - 2(2) = -5$$
$$D = -5$$

Step 2:　Solve for Dx.

right side of original equations

second column of D matrix—values of y

$$Dx = \begin{bmatrix} 8 & 2 \\ 6 & -1 \end{bmatrix} = 8(-1) - 2(6) = -20$$
$$Dx = -20$$

Step 3:　Solve for Dy.

First column of D matrix—values of x

$$Dy = \begin{bmatrix} 1 & 8 \\ 2 & 6 \end{bmatrix} = 1(6) - 8(2) = -10$$
$$Dy = -10$$

Step 4:

$$x = \frac{Dx}{D} = \frac{-20}{-5} = 4$$
$$y = \frac{Dy}{D} = \frac{-10}{-5} = 2$$

Solution of the equation is (4, 2).

Proof:
Equation 1:

$$4 + 2(2) = 8$$

Equation 2:

$$2(4) - 2 = 6$$

Again, this may seem like a slow way to solve a problem, but the computer can produce thousands of such solutions very quickly after a simple program has been developed. The actual programming aspects are beyond the scope of this text, but it is pretty obvious that if a large volume of problems are to be solved, it would be very time consuming and uneconomical to do the work in any way that did not include the use of computers.

It is necessary for the programmer to understand the method of solution so that he can translate the requirements of the job into computer language. Following is an example of a flow chart and method of solution of simultaneous linear equations by computer processing, using matrices.

Problem:

Solve for x and y (as in the following):

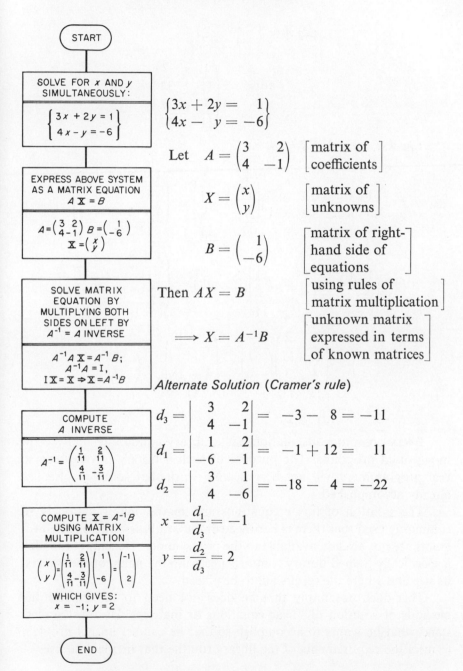

$$\begin{cases} 3x + 2y = 1 \\ 4x - y = -6 \end{cases}$$

Let $A = \begin{pmatrix} 3 & 2 \\ 4 & -1 \end{pmatrix}$ $\begin{bmatrix} \text{matrix of} \\ \text{coefficients} \end{bmatrix}$

$X = \begin{pmatrix} x \\ y \end{pmatrix}$ $\begin{bmatrix} \text{matrix of} \\ \text{unknowns} \end{bmatrix}$

$B = \begin{pmatrix} 1 \\ -6 \end{pmatrix}$ $\begin{bmatrix} \text{matrix of right-} \\ \text{hand side of} \\ \text{equations} \end{bmatrix}$

Then $AX = B$ $\begin{bmatrix} \text{using rules of} \\ \text{matrix multiplication} \end{bmatrix}$

$\implies X = A^{-1}B$ $\begin{bmatrix} \text{unknown matrix} \\ \text{expressed in terms} \\ \text{of known matrices} \end{bmatrix}$

Alternate Solution (Cramer's rule)

$d_3 = \begin{vmatrix} 3 & 2 \\ 4 & -1 \end{vmatrix} = -3 - 8 = -11$

$d_1 = \begin{vmatrix} 1 & 2 \\ -6 & -1 \end{vmatrix} = -1 + 12 = 11$

$d_2 = \begin{vmatrix} 3 & 1 \\ 4 & -6 \end{vmatrix} = -18 - 4 = -22$

$x = \dfrac{d_1}{d_3} = -1$

$y = \dfrac{d_2}{d_3} = 2$

Matrix Solution:

$$|A| = -3 - 8 = -11$$

$$\text{adj } A = \begin{pmatrix} -1 & -2 \\ -4 & 3 \end{pmatrix}$$

$$A^{-1} = \frac{\text{adj } A}{A} = \begin{pmatrix} \dfrac{1}{11} & \dfrac{2}{11} \\ \dfrac{4}{11} & -\dfrac{3}{11} \end{pmatrix}$$

Checks: $AA^{-1} = I = A^{-1}A$

$$I = \begin{pmatrix} 1 & 0 \\ 0 & 1 \end{pmatrix} = \begin{pmatrix} 3 & 2 \\ 4 & -1 \end{pmatrix} \begin{pmatrix} \dfrac{1}{11} & \dfrac{2}{11} \\ \dfrac{4}{11} & -\dfrac{3}{11} \end{pmatrix} = \begin{pmatrix} \dfrac{3+8}{11} & \dfrac{6-6}{11} \\ \dfrac{4-4}{11} & \dfrac{8+3}{11} \end{pmatrix}$$

$$I = \begin{pmatrix} 1 & 0 \\ 0 & 1 \end{pmatrix} = \begin{pmatrix} \dfrac{1}{11} & \dfrac{2}{11} \\ \dfrac{4}{11} & -\dfrac{3}{11} \end{pmatrix} \begin{pmatrix} 3 & 2 \\ 4 & -1 \end{pmatrix} = \begin{pmatrix} \dfrac{3+8}{11} & \dfrac{2-2}{11} \\ \dfrac{12-12}{11} & \dfrac{8+3}{11} \end{pmatrix}$$

$$\Longrightarrow \begin{pmatrix} x \\ y \end{pmatrix} = \begin{pmatrix} \dfrac{1}{11} & \dfrac{2}{11} \\ \dfrac{4}{11} & \dfrac{3}{11} \end{pmatrix} \begin{pmatrix} 1 \\ -6 \end{pmatrix} = \begin{pmatrix} \dfrac{1-12}{11} \\ \dfrac{4+18}{11} \end{pmatrix} = \begin{pmatrix} -1 \\ 2 \end{pmatrix}$$

$$\Longrightarrow x = -1, y = 2$$

Every computer installation has a library containing many commonly used programs. The purpose of such a library is obvious—so that programmers will not be repeating work that someone else has already accomplished.

The solution of linear equations and matrix manipulation is so commonly used that there is hardly a computer library where such programs are not already available to the programmer. He simply calls for a specifically named library routine, specifying the name of the matrix, its size and where the results are to be placed.

This does not imply that he does not need to understand the methods of solution of linear equations or matrices. He must understand what he wants to accomplish so that he can set up his problem to meet the requirements of the library routine that he plans to use.

REVIEW QUESTIONS

7.22. For the following matrix, write the formula to solve for the determinant.

$$A = \begin{bmatrix} a_{11} & a_{12} \\ a_{21} & a_{22} \end{bmatrix}$$

7.23. Define the following terms with respect to the solution of simultaneous linear equations.
(a) Inconsistent
(b) Dependent

7.24. Show the two equations to be used when solving for two variables (x and y).

7.25. Solve the following equations with the determinant method. Show the three matrices developed from this problem.
Equation 1: $2x + y = 10$
Equation 2: $x - 2y = 11$

Solving for Three Variables

When systems of linear equations are to be solved for three variables, there are a number of ways to approach a solution. The method we will discuss is usually called Cramer's rule.

The first step is to compute the determinant of the basic matrix. This involves a little more arithmetic than did the 2×2 matrix, but the idea is much the same.

Step 1:

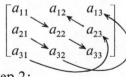

(a) Multiply the main diagonal elements (a_{11}, a_{22}, a_{33}).
(b) Add the multiplied elements (a_{21}, a_{32}, a_{13}).
(c) Add the multiplied elements (a_{31}, a_{23}, a_{12}).

Step 2:

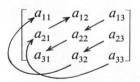

(a) Subtract the multiplied opposite diagonal (a_{13}, a_{22}, a_{31}).
(b) Subtract the multiplied elements (a_{23}, a_{32}, a_{11}).
(c) Subtract the multiplied elements (a_{33}, a_{21}, a_{12}).

$$D = (a_{11})(a_{22})(a_{33}) + (a_{21})(a_{32})(a_{13}) + (a_{31})(a_{23})(a_{12})$$
$$- (a_{13})(a_{22})(a_{31}) - (a_{23})(a_{32})(a_{11}) - (a_{33})(a_{21})(a_{12})$$

The determinant of any set of three simultaneous equations may be derived with this technique.

When the determinant (D) has been found, x, y, and z are solved in much the same manner. The matrix is changed for each variable, replacing the numbers associated with that variable by the right-hand sides of the three equations.

The actual computation of the variables is accomplished with the following equations:

$$x = \frac{Dx}{D} \qquad \text{determinant of matrix solving for } Dx$$
$$\phantom{x = \frac{Dx}{D} \qquad} \text{determinant of the original system}$$

$$y = \frac{Dy}{D}$$

$$z = \frac{Dz}{D}$$

An annotated example should help to clarify the method.

EXAMPLE

Equation 1:

$$x + y + 2z = 8$$

Equation 2:

$$2x - 3y + z = 1$$

Equation 3:

$$x + 2y - z = -7$$

$$D = \begin{bmatrix} 1 & 1 & 2 \\ 2 & -3 & 1 \\ 1 & 2 & -1 \end{bmatrix}$$

Before solving for D, we will set up the matrices for the determinants that will be used to solve for x, y, and z.

$$Dx = \begin{bmatrix} 8 & 1 & 2 \\ 1 & -3 & 1 \\ -7 & ? & -1 \end{bmatrix} \qquad x = \frac{Dx}{D}$$

Note the replacement of the "x" variable items.

$$Dy = \begin{bmatrix} 1 & 8 & 2 \\ 2 & 1 & 1 \\ 1 & -7 & -1 \end{bmatrix} \qquad y = \frac{Dy}{D}$$

Now the "y" variables are replaced.

$$Dz = \begin{bmatrix} 1 & 1 & 8 \\ 2 & -3 & 1 \\ 1 & 2 & -7 \end{bmatrix} \qquad z = \frac{Dz}{D}$$

Finally, the "z" variables are replaced.

Each matrix is solved for its determinant, followed by the final division steps to arrive at $x, y,$ and z.

$$\begin{aligned} |D| = {}& (1)(-3)(-1) + (2)(2)(2) + (1)(1)(1) - (2)(-3)(1) \\ & - (1)(2)(-1) - (1)(2)(1) \\ & 3 + 8 + 1 + 6 + 2 - 2 = 18 \\ |A| = {}& 8(-3)(-1) + (1)(2)(2) + (-7)(1)(1) - (-7)(-3)(2) \\ & - (1)(1)(-1) - (8)(2)(1) \\ & 24 + 4 - 7 - 42 + 1 - 16 = -36 \\ & x = \frac{-36}{18} = -2 \end{aligned}$$

$|B|$ and $|C|$ are solved in the same manner.

REVIEW QUESTIONS

7.26. Solve for Dy and Dz in the example above. The final solution will be $(-2, ?, ?)$.

7.27. (a) Set up the basic matrices to solve the following equations:
  Equation 1: $x - y + z = 3$
  Equation 2: $3x + 2y - z = 0$
  Equation 3: $2x + y + 2z = 3$
  (b) Solve for x, y and z.

SPECIAL MATRICES

There are a number of special matrix techniques and special types of matrices with which the student should be familiar to the extent that he realizes that they exist. No attempt will be made at this time to provide practical uses for these matrices, but the knowledge of them will be available when needed.

Transpose

The *transpose* of a matrix is accomplished by interchanging the rows and columns of the matrix.

EXAMPLES

1.
$$A = \begin{bmatrix} 2 & 3 & -2 \\ 1 & -2 & -4 \end{bmatrix}$$

Transpose is signified with the superscript T.

$$A^T = \begin{bmatrix} 2 & 1 \\ 3 & -2 \\ -2 & -4 \end{bmatrix}$$

2.
$$A = \begin{bmatrix} 2 & -1 \\ -3 & 4 \end{bmatrix} \qquad A^T = \begin{bmatrix} 2 & -3 \\ -1 & 4 \end{bmatrix}$$

Note that a square matrix stays square, but the 2×3 matrix changed to a 3×2 matrix.

Three rules can be stated for the transpose of a matrix.

Rule 1: The transpose of a transposed matrix will be the original matrix.

$$(A^T)^T = A$$

Rule 2: The transpose of the sum of two matrices (A and B) equals the transpose of A plus the transpose of B.

$$(A + B)^T = A^T + B^T$$

Rule 3: The transpose of the multiplication of two matrices (A and B) equals the transpose of B times the transpose of A.

$$(AB)^T = B^T A^T$$

Rules 2 and 3 are applicable only if the matrices conform to the rules for matrix addition and multiplication.

Symmetric Matrices

A matrix is *symmetric* when the transpose is identical to the original matrix.

EXAMPLE

$$A = \begin{bmatrix} 2 & -3 & 1 \\ -3 & -1 & -2 \\ 1 & -2 & 3 \end{bmatrix} \qquad A^T = \begin{bmatrix} 2 & -3 & 1 \\ -3 & -1 & -2 \\ 1 & -2 & 3 \end{bmatrix}$$

Symmetric matrices must be square. The example above shows that

$$A = A^T$$

Rule 4: If a matrix and its transpose are multiplied, the product will be symmetric.

$$AA^T = B \text{ (symmetric matrix)}$$

Proof:

1.
$$A = \begin{bmatrix} 2 & 3 & -2 \\ 1 & -2 & -4 \end{bmatrix} \qquad A^T = \begin{bmatrix} 2 & 1 \\ 3 & -2 \\ -2 & -4 \end{bmatrix}$$

Multiplying the two matrices results in

$$AA^T = \begin{bmatrix} 17 & 4 \\ 4 & 21 \end{bmatrix} \qquad (AA^T)^T = \begin{bmatrix} 17 & 4 \\ 4 & 21 \end{bmatrix}$$

2.
$$A = \begin{bmatrix} 2 & 1 & -1 \\ 3 & -2 & 1 \\ 1 & -2 & 2 \\ -3 & 2 & 1 \end{bmatrix} \qquad A^T = \begin{bmatrix} 2 & 3 & 1 & -3 \\ 1 & -2 & -2 & 2 \\ -1 & 1 & 2 & 1 \end{bmatrix}$$

A 4 × 3 matrix multiplied by a 3 × 4 matrix will result in a 4 × 4 matrix:

$$AA^T = \begin{bmatrix} 6 & 3 & -2 & -5 \\ 3 & 14 & 9 & -12 \\ -2 & 9 & 9 & -5 \\ -5 & -12 & -5 & 14 \end{bmatrix}$$

$$(AA^T)^T = \begin{bmatrix} 6 & 3 & -2 & -5 \\ 3 & 14 & 9 & -12 \\ -2 & 9 & 9 & -5 \\ -5 & -12 & -5 & 14 \end{bmatrix}$$

Triangular and Diagonal Matrices

A *triangular* matrix is a square matrix that contains all zeros either above or below the main diagonal. The diagonal itself may either contain real numbers or may also contain zeros.

EXAMPLES

$$A = \begin{bmatrix} 1 & 3 & -2 \\ 0 & 2 & -1 \\ 0 & 0 & 3 \end{bmatrix} \quad B = \begin{bmatrix} 0 & 0 & 0 \\ 1 & 2 & 0 \\ 2 & 3 & 0 \end{bmatrix}$$

$$C = \begin{bmatrix} 0 & 0 & 0 \\ 0 & 0 & 0 \\ 0 & 0 & 0 \end{bmatrix}$$

If all elements *not* on the main diagonal of a matrix are zeros, the matrix is called *diagonal*. Elements on the main diagonal may be zeros or other numbers.

EXAMPLES

$$A = \begin{bmatrix} 1 & 0 & 0 \\ 0 & 0 & 0 \\ 0 & 0 & 0 \end{bmatrix} \quad B = \begin{bmatrix} 1 & 0 & 0 \\ 0 & 3 & 0 \\ 0 & 0 & 2 \end{bmatrix}$$

Note that *diagonal* matrices are also *triangular*. See matrix C above.

Submatrices

It is possible to make a number of square submatrices from a single rectangular matrix.

EXAMPLES

1. $\qquad A = \begin{bmatrix} 1 & 2 & 3 \\ 4 & 5 & 6 \end{bmatrix}$ (2 × 3 matrix)

A 2 × 3 matrix can be subdivided into three different square matrices:

$$B = \begin{bmatrix} 1 & 2 \\ 4 & 5 \end{bmatrix} \quad C = \begin{bmatrix} 2 & 3 \\ 5 & 6 \end{bmatrix} \quad D = \begin{bmatrix} 1 & 3 \\ 4 & 6 \end{bmatrix}$$

2. $\qquad A = \begin{bmatrix} 1 & 2 & 3 & 4 \\ 5 & 6 & 7 & 8 \\ 9 & 10 & 11 & 12 \end{bmatrix}$ (3 × 4 matrix)

Subdivided:

$$B = \begin{bmatrix} 1 & 2 & 3 \\ 5 & 6 & 7 \\ 9 & 10 & 11 \end{bmatrix} \quad C = \begin{bmatrix} 2 & 3 & 4 \\ 6 & 7 & 8 \\ 10 & 11 & 12 \end{bmatrix}$$

$$D = \begin{bmatrix} 1 & 2 & 4 \\ 5 & 6 & 8 \\ 9 & 10 & 12 \end{bmatrix} \quad E = \begin{bmatrix} 1 & 3 & 4 \\ 5 & 7 & 8 \\ 9 & 11 & 12 \end{bmatrix}$$

(plus a whole series of 2×2 matrices).

This concept is often useful since certain operations can be performed on square matrices that cannot be performed on rectangular matrices.

REVIEW QUESTIONS

7.28. Show the transpose of the following matrices.

(a) $A = \begin{bmatrix} 3 & -2 & 2 & -1 \\ -1 & 2 & -3 & -2 \\ -2 & 3 & 4 & 1 \end{bmatrix}$ (b) $B = \begin{bmatrix} 2 & 3 \\ 4 & 5 \\ 6 & 7 \end{bmatrix}$

7.29. What is the rule for a symmetric matrix?

7.30. What type of matrix is each of the following matrices?

(a) $A = \begin{bmatrix} 1 & 2 & -3 \\ 0 & 0 & -2 \\ 0 & 0 & 3 \end{bmatrix}$ (b) $B = \begin{bmatrix} 2 & 0 & 0 \\ 0 & 0 & 0 \\ 0 & 0 & -1 \end{bmatrix}$

7.31. How many submatrices can be made from each of the following matrices?

(a) $A = \begin{bmatrix} 2 & -1 & 3 \\ -2 & 4 & 1 \end{bmatrix}$ (b) $B = \begin{bmatrix} -1 & 2 & 3 \\ -2 & -3 & 1 \\ 3 & 2 & -2 \end{bmatrix}$

REVIEW

Review of Symbols

D	determinant
$|A|$	determinant of matrix A
A^T	transpose of matrix A

Review of Terminology

Matrix	an array of numbers or symbols, arranged in a sequence of rows and columns
Size (of a matrix)	the number of rows and columns in a matrix
Square matrix	a matrix which contains the same number of rows as columns
Main diagonal	in a square matrix, the elements from the upper left corner through the lower right corner
Equal matrices	two matrices in which the corresponding elements are identical
Submatrices	smaller sized square matrices made from a single rectangular matrix
Equivalent matrices	two matrices that are the same size, i.e., have the same number of rows and columns
Vector	a single row matrix or a single column matrix
Determinant	the single value of a square matrix obtained from the sum of the products formed in accordance with a specific set of rules
Inconsistent	no solution to matrix possible as there will be no point of intersection
Dependent	no solution to matrix possible as there will be an infinite number of intersections
Transpose	to interchange rows and columns of a matrix
Symmetric matrix	one in which the transpose is identical to the original matrix
Triangular matrix	a square matrix containing all zeros either above or below the main diagonal
Diagonal matrix	a square matrix with all elements not on the main diagonal being zeros

8 LINEAR PROGRAMMING

INTRODUCTION

Linear programming is a mathematical method for solving problems that are usually considered to be business or tactically oriented rather than scientifically oriented.

New methods of approaching the solutions to problems, called *operations research*, were developed during World War II to solve various military tactical problems. The overall method used was to assemble a team of experts (physicists, biologists, mathematicians, psychologists, etc.) and then attack the problem—covering every aspect with an expert in that particular area of knowledge.

These methods of problem solving worked so well that the technique of operations research (OR) moved into the field of big business after the war was over and OR companies are still flourishing today. One of the techniques used by OR technicians is the mathematical method of solving nonmathematical problems, called linear programming.

The major facets of the technique are the *maximizing* of *profits* and the *minimizing* of *costs*. Since these topics are extremely important to all businesses, it is no wonder that interest in this field is always active. This chapter will explore some basic techniques of linear programming.

MAXIMIZING PROFITS

The best way to approach this topic is with an example.

EXAMPLE

Company *B* is a manufacturer of watches. It makes two models which are produced in two shops, some of the work on each model being performed in both shops. The expensive model (called the Imperial) wholesales at $50.00 with a profit of $25.00 and the inexpensive watch (called the Standard) wholesales at $25.00 with a profit of $10.00.

Shop *A* has 320 hours a week of working time available for producing Imperial (*I*) and Standard (*S*) watches. Shop *B* has 240 hours of working time for the same purpose.

It takes 20 hours to produce an *I* watch and 8 hours to produce an *S* watch. Every *I* watch gets 12 hours of work in shop *A* and 8 hours in shop *B*. Every *S* watch gets 3 hours of work in shop *A* and 5 hours in shop *B*.

What would be the best production mix of *I* and *S* to provide the greatest profit to the company?

Let P = Total profit
I = Imperial watch
S = Standard watch

Estimation Method of Solution

The first step is to examine the known facts and to organize them into usable form.

What do we know about profit? There is $25.00 profit on *I* and $10.00 profit on *S*. The equation can then be written:

$$P = 25I + 10S$$

Also, there are certain restrictions: Shop *A* cannot exceed 320 hours and shop B cannot exceed 240 hours; it takes 12 hours for I and 3 hours for S in shop A and 8 hours for I and 5 hours for S in shop B. These restrictions can be written:

$$\text{Shop } A: \quad 12I + 3S \leqslant 320$$
$$\text{Shop } B: \quad 8I + 5S \leqslant 240$$

The first possibility that comes to mind is, "What if only I watches were produced?" We know that there is a total of 560 hours available in the two shops $(320 + 240)$ and we know that it takes a total of 20 hours to produce one I, so we can quickly get a gross figure:

$$\frac{560}{20} = 28I \text{ (maximum that can be produced)}$$

then

$$P_I = 28 \times 25 \text{ (profit on each } I) = \$700.00$$

For clarity, we will set this last equation in vector form and use matrix multiplication:

$$P_I = (28 \quad 0)\begin{pmatrix} I25 \\ S10 \end{pmatrix} = 700 + 0 = \$700.00$$

number of watches made

profit

Now, we must check to make sure that the restrictions have not been exceeded. The manufacturing time for each model in each shop is

$$\text{Shop } A: \quad \begin{matrix} I \\ S \end{matrix}\begin{pmatrix} 12 \\ 3 \end{pmatrix} \qquad \text{Shop } B: \quad \begin{matrix} I \\ S \end{matrix}\begin{pmatrix} 8 \\ 5 \end{pmatrix}$$

Checking for I only, we have

$$\text{Shop } A: \quad (28 \quad 0)\begin{pmatrix} 12 \\ 0 \end{pmatrix} = 336 \qquad \text{(320 exceeded—28 } I\text{'s cannot be produced)}$$

$$\text{Shop } B: \quad (28 \quad 0)\begin{pmatrix} 8 \\ 0 \end{pmatrix} \quad 224 \qquad \text{(240 not exceeded—OK)}$$

The limits of shop A have been exceeded by 16 hours. If we drop the production by two I's, we will be within limits:

$$\text{Shop } A: \quad (26 \quad 0)\begin{pmatrix} 12 \\ 0 \end{pmatrix} = 312 \qquad \text{(320 not exceeded—OK)}$$

$$\text{Shop } B: \quad (26 \quad 0)\begin{pmatrix} 8 \\ 0 \end{pmatrix} = 208 \qquad \text{(240 not exceeded—OK)}$$

Now, the adjusted profit:

$$P_I = (26 \quad 0)\binom{25}{10} = 650 + 0 = \$650.00$$

The same procedure may be followed to determine the profit if all S watches were produced.

$$\frac{560}{8} = 70S \qquad P = 70 \times 10 = \$700.00$$

$$(0 \quad 70)\binom{25}{10} = 0 + 700 = \$700.00$$

Check restrictions:

Shop A: $(0 \quad 70)\binom{0}{3} = 210$ (320 not exceeded—OK)

Shop B: $(0 \quad 70)\binom{0}{5} = 350$ (240 exceeded—70S's cannot be produced)

How many times 5 hours do we have to take away from 350 to reach the 240 limit required?

$$\begin{array}{r} 350 \\ -240 \\ \hline 110 \text{ hours to lose} \end{array} \qquad 22 \text{ watches} \quad 5\overline{)110} \qquad \begin{array}{r} 70 \\ -22 \\ \hline 48 \text{ maximum } S\text{'s} \end{array}$$

Check restrictions again:

Shop A: $(0 \quad 48)\binom{0}{3} = 144$ ($<$320)

Shop B: $(0 \quad 48)\binom{0}{5} = 240$ ($=$240)

Now shop B does not exceed the restriction, but shop A is left practically without work; therefore, it is not practical to produce just S watches.

To this point, we have determined that:

1. It is not feasible to produce just S watches.

2. It is possible to produce just I watches, yielding a profit of $650.00 and leaving slack time of 8 hours in shop A and 32 hours in shop B, or a total of 40 hours of slack time.

Just for the exercise, let us pick two arbitrary numbers and see how it works out—20I and 20S. Check restrictions first:

Shop A: $(20 \quad 20)\binom{12}{3} = 240 + 60 = 300$ ($<$320—OK)

Shop B: $(20 \quad 20)\binom{8}{5} = 160 + 100 = 260$ ($<$240—not OK)

That is only 20 hours too much. We could drop two I's and one S to bring it down below the required level:

Shop A: $(18 \quad 19)\binom{12}{3} = 216 + 57 = 273 \quad (<320—\text{OK})$

Shop B: $(18 \quad 19)\binom{8}{5} = 144 + 95 = 239 \quad (<240—\text{OK})$

We have met the restrictions, so let us figure the profit.

$$P = (18 \quad 19)\binom{25}{10} = 450 + 190 = \$640.00$$

How does this compare with the production of all I's?
Profit: $640.00
Slack time: Shop A—47 hours
Shop B— 1 hour
This mix produces $10.00 less profit and 8 hours more slack time than producing just I watches.

To eliminate the excess 20 hours in shop B, we could have used $20I$ and $16S$ instead of $18I$ and $19S$. This would reduce the original $20I$, $20S$ figure by 20 hours (four 5's).

Shop A: $(20 \quad 16)\binom{12}{3} = 240 + 48 = 288 \quad (<320 - \text{OK})$

Shop B: $(20 \quad 16)\binom{8}{5} = 160 + 80 = 240 \quad (=240 - \text{OK})$

$$P = (20 \quad 16)\binom{25}{10} = 500 + 160 = \$660.00$$

This results in $10.00 more profit and much less slack time in the shops than producing just I watches.

The final recommendation would be: The best procedure for Company B would be to produce $20I$ and $16S$ watches ($660.00 profit and 32 hours slack time). Additional improvement could be made if some method were devised to eliminate some of the excessive slack time in shop A (32 hours for mixed production).

REVIEW QUESTIONS

8.1. Name two business uses of the technique called linear programming.

8.2. What is the name of the problem solving system that uses linear programming as one of its techniques?

8.3. The restrictions on Company B in the previous example were specified to be:

$$\text{Shop A:} \quad 12I + 3S \leqslant 320$$
$$\text{Shop B:} \quad 8I + 5S \leqslant 240$$

(a) Assume that one additional technician was hired for each shop to work a normal 8-hour day, 5-day week. How would that change the restrictions?

(b) Assume that, with the aid of time and motion studies, two hours of production time were cut in each shop for I watches and one hour was cut in shop B for the production of S watches. How would that change the restrictions?

(c) With these new restrictions, recompute:
   (1) the maximum profit for producing just I. How much slack time in each shop? How many I's produced?
   (2) the maximum profit for producing just S. How much slack in each shop? How many S's produced?
   (3) the best possible mix of I and S to provide the greatest profit to the company. How much slack time in each shop?

Algebraic Method of Solution

Considering the example on the previous pages, we can change the restrictions into equations in the following manner: The restriction on shop A is 320 hours. Whenever less time than 320 hours is used, the shop is not used to its full potential and this slack time can be called the *slack variable*. The restrictions changed into equation form would be:

Equation 1: (Shop A) $12I + 3S + x_1 = 320$ (where x_1 is the slack time)

Equation 2: (Shop B) $8I + 5S + x_2 = 240$ (x_2 is slack time)

Add to the above two equations the basic profit equation:

Equation 3: $25I + 10S = P$

The process is to keep manipulating the equations so that each time they are manipulated the profit increases, until it reaches the maximum profit limit without exceeding the restrictions. A start can be made at the point of zero profit:

$$I = 0 \quad\quad S = 0 \quad\quad x_1 = 320 \quad\quad x_2 = 240 \quad\quad P = 0$$

This implies that no watches were produced; all of the shop time was slack time and zero profit resulted. This is obviously the lowest point on the profit scale that can be reached.

The next step is to choose a *key variable*, preferably the one with the largest profit amount (I in equation 3). At the same time, x_1 and the other variable (S) are set to zero. Now we must determine the largest possible value of the key variable I. Dividing 320 by 12 results in 26+. As it is not feasible to produce part of a watch, we will drop the fraction.

The next step is to decide on the *pivotal equation*. This will be the equation with the smallest quotient to avoid running into negative numbers. In equation 2, dividing 240 by 8 results in a quotient of 30; therefore, equation 1 (with a quotient of 26) becomes the pivotal equation.

These preliminary steps are very important and must be carried out in a very careful manner or the results will not be the optimum for profit.

Divide each side of the pivotal equation by the coefficient (actual number) of the key variable (I).

Original equation: $\qquad 12I + 3S + x_1 = 320$

Changed equation: $\qquad I + \dfrac{3}{12}S + \dfrac{1}{12}x_1 = 26$

Since both S and x_1 are zero, the result is $I = 26$.

$$I + \frac{3}{12} \times 0 + \frac{1}{12} \times 0 = 26$$

Note that this is the same result that was obtained in the estimation method. The profit result will also be the same:

$$P_I = 26 \times 25 = \$650.00$$

With additional manipulation of equations, the final optimum result can be reached, but this is a very difficult and hard-to-understand technique, particularly for nonmathematicians.

For problems that do not have too many variables or too many restrictions, the estimation method can prove fairly easy to use. When the number of variables or restrictions becomes excessive, it is usually possible to partition the overall problem into a number of subproblems, which can then be solved easily.

The estimation method takes a certain amount of trial-and-error steps and mathematicians much prefer the algebraic method (although it also requires a considerable number of trial-and-error steps). On the other hand, nonmathematicians can readily grasp the techniques of the estimation method and they invariably flounder with the algebraic method. An example of the algebraic method is shown below for those who may be interested in the technique.

EXAMPLE

Problem Statement

A fruit orchard presently contains 20 fruit trees per acre, and the average yield is 600 pieces of fruit per tree. For each additional tree planted, the average yield per tree is reduced by 15 pieces of fruit per tree. How many trees per acre will result in the largest fruit crop?

Solution:

Let

$$x = \text{quantity of new trees planted per acre}$$

Then

$$20 + x = \text{quantity of trees per acre}$$
$$600 - 15x = \text{average quantity of fruit per tree}$$
$$y(x) = \text{total fruit yield}$$
$$= \text{(quantity of trees) (fruit yield per tree)}$$
$$= (20 + x)(600 - 15x)$$
$$= 12{,}000 + 300x - 15x^2$$
$$y'(x) = 300 - 30x \Big\rbrace$$
$$y''(x) = -30 < 0$$

Necessary conditions for maximum of $y(x)$:

Find x_0 such that $y'(x_0) = 0$

Compute $y_z = y''(x_0)$

If $y_z < 0$ then $y_0 = y(x_0)$ is maximum y value

$$\therefore \quad 300 - 30x_0 = 0 \longrightarrow x_0 = 10$$
$$\therefore \quad y(x_0) = (30)(450) = 13{,}500$$

$\longrightarrow$ 10 new trees planted will result in the maximum yield of 13,500 pieces of fruit.

Computation:

$$y = 12{,}000 + 300x - 15x^2$$
$$y + \Delta y = 12{,}000 + 300(x + \Delta x) - 15(x + \Delta x)^2$$
$$= 12{,}000 + 300x + 300(\Delta x) - 15x^2 - 30x(\Delta x) - 15(\Delta x)^2$$
$$\Delta y = 300(\Delta x) - 30x(\Delta x) - 15(\Delta x)^2$$

$$\frac{\Delta y}{\Delta x} = 300 - 30x - 15(\Delta x) \qquad (\Delta x \neq 0)$$

$$y' \equiv \lim_{\Delta x \to 0} \left(\frac{\Delta y}{\Delta x}\right) = 300 - 30x$$

$$y'' \equiv (y')'$$

$$y' + \Delta y' = 300 - 30(x + \Delta x) = 300 - 30x - 30(\Delta x)$$

$$\Delta y' = -30(\Delta x)$$

$$\frac{\Delta y'}{\Delta x} = -30$$

$$y'' = \lim_{\Delta x \to 0} \frac{\Delta y'}{\Delta x} = -30$$

Now, let us try the estimation method on the same problem.

1. We know that at present there are 20 trees per acre which yields 600 pieces of fruit per tree. What is the present yield?

$$(20)(600) = 12,000 \text{ pieces of fruit}$$

---

2. We want to add trees until we reach the maximum crop size. Each extra tree planted will reduce the yield by 15 pieces of fruit. Let x = added trees, $y(x)$ = crop size.

$$(20 + x)(600 - 15x) = y(x)$$

(20 trees + "x" trees times 600 pieces minus 15 pieces for each added tree will give the new crop size)

---

3. Try the addition of five trees per acre to see what happens:

$$(20 + 5)(600 - (15)(5)) =$$
$$(25)(525) = 13,125$$

---

4. Adding five trees increases the crop. Try 10 trees:

$$(20 + 10)(600 - (15)(10)) =$$
$$(30)(450) = 13,500$$

---

5. This again increased the yield. Try 11 trees:

$$(20 + 11)(600 - (15)(11)) =$$
$$(31)(435) = 13,485$$

At 11 trees, the yield dropped, so 10 trees reach the maximum yield of 13,500 pieces of fruit. The answer is identical to the algebraic method shown on the previous pages.

REVIEW QUESTIONS

8.4. Change the ground rules for the fruit orchard problem: 25 trees per acre with average yield of 250 pieces of fruit per tree. Each added tree reduces the yield by 20 pieces of fruit. Each tree taken away adds 20 pieces to the yield.

(a) What is the present fruit yield?

(b) Show the equations to compute maximum crop size (both by adding to and by taking away from the number of trees).

(c) What is the maximum crop size possible?

(d) How many trees per acre should be added or taken away to attain the maximum crop size?

Example With More Variables

Whatever the method of solution, the more variables there are the more difficult the problem becomes. The problem that follows is an example of the increase in difficulty.

Problem Statement

Company *H* produces three different types of radio sets. The following information is available:

	Hours to Produce	Sale Price	Cost	Profit
Model *A*	10	$50.00	$35.00	$15.00
Model *B*	8	$40.00	$28.00	$12.00
Model *C*	6	$30.00	$21.00	$ 9.00

Two shops produce the three models:

	Manufacturing Time	
	Shop X	Shop Y
Model *A*	4	6
Model *B*	3	5
Model *C*	4	2
Total working time available	280	360

The following profit equation and restrictions are based on the above information:

$$P = 15A + 12B + 9C$$

Restrictions:

$$\text{Shop } X: \quad 4A + 3B + 4C \leqslant 280$$
$$\text{Shop } Y: \quad 6A + 5B + 2C \leqslant 360$$

If only A's produced:

$$\frac{640}{10} = 64 \qquad P_A = (64 \quad 0 \quad 0)\begin{pmatrix} 15 \\ 12 \\ 9 \end{pmatrix} = 960 + 0 + 0 = \$960.00$$

Check restrictions:

$$\text{Shop } X: \quad (4 \quad 0 \quad 0)\begin{pmatrix} 64 \\ 0 \\ 0 \end{pmatrix} = 256 \quad (<280\text{—OK})$$

$$\text{Shop } Y: \quad (6 \quad 0 \quad 0)\begin{pmatrix} 64 \\ 0 \\ 0 \end{pmatrix} = 384 \quad (>360\text{—not OK}) \text{ must reduce}$$
$$\text{by 26 hours,}$$
$$\text{5 radios}$$

Recompute restrictions:

$$\text{Shop } X: \quad (4 \quad 0 \quad 0)\begin{pmatrix} 59 \\ 0 \\ 0 \end{pmatrix} = 236 \quad (<280\text{—OK})$$

$$\text{Shop } Y: \quad (6 \quad 0 \quad 0)\begin{pmatrix} 59 \\ 0 \\ 0 \end{pmatrix} = 354 \quad (<360\text{—OK})$$

59 A's produced—
Slack time: Shop X—44 hours
Shop Y—6 hours

$$P_A = (59 \quad 0 \quad 0)\begin{pmatrix} 15 \\ 12 \\ 9 \end{pmatrix} = 885 + 0 + 0 = \$885.00$$

If only B's produced:

$$\frac{640}{8} = 80 \qquad P_B = (0 \quad 80 \quad 0)\begin{pmatrix} 15 \\ 12 \\ 9 \end{pmatrix} = 0 + 960 + 0 = \$960.00$$

Check restrictions:

$$\text{Shop } X: \quad (0 \quad 3 \quad 0)\begin{pmatrix} 0 \\ 80 \\ 0 \end{pmatrix} = 0 + 240 + 0 = 240 \quad (<280\text{—OK})$$

$$\text{Shop } Y: \quad (0 \quad 5 \quad 0)\begin{pmatrix} 0 \\ 80 \\ 0 \end{pmatrix} = 0 + 400 + 0 = 400 \quad (>360\text{—not OK}) \text{ must reduce by 40 hours, 8 radios}$$

Recompute restrictions:

$$\text{Shop } X: \quad (0 \quad 3 \quad 0)\begin{pmatrix} 0 \\ 72 \\ 0 \end{pmatrix} = 0 + 276 + 0 = 276 \quad (<280\text{—OK})$$

$$\text{Shop } Y: \quad (0 \quad 5 \quad 0)\begin{pmatrix} 0 \\ 72 \\ 0 \end{pmatrix} = 0 + 360 + 0 = 360 \quad (=360\text{—OK})$$

72 B's produced

Slack time: Shop X—4 hours
　　　　　　Shop Y—0 hours

$$P_B = (0 \quad 72 \quad 0)\begin{pmatrix} 15 \\ 12 \\ 9 \end{pmatrix} = 0 + 864 + 0 = \$864.00$$

If only C's produced:

$$\frac{640}{6} = 106 + (\text{drop fraction})$$

$$P_C = (0 \quad 0 \quad 106)\begin{pmatrix} 15 \\ 12 \\ 9 \end{pmatrix} = 0 + 0 + 954 = \$954.00$$

Check restrictions:

$$\text{Shop } X: \quad (0 \quad 0 \quad 4)\begin{pmatrix} 0 \\ 0 \\ 106 \end{pmatrix} = 0 + 0 + 424 \quad (>280\text{—not OK})$$

$$\text{Shop } Y: \quad (0 \quad 0 \quad 2)\begin{pmatrix} 0 \\ 0 \\ 106 \end{pmatrix} = 0 + 0 + 212 \quad (<360\text{—OK})$$

The time in shop X has been exceeded by so much that it is obvious that further calculation would be fruitless; it is not feasible to produce just C's.

The highest profit to this point is production of just A's = \$885.00. What mix of A, B, and C will produce the greatest profit and still keep the shops within the specified restrictions?

1. Since A has the greatest profit, try a combination with a little emphasis on A; then immediately check restrictions to see how far off the first estimation runs.

$$\text{Shop } X: \quad (29 \quad 37 \quad 0)\begin{pmatrix}4\\3\\4\end{pmatrix} = 116 + 111 + 0 = 227 \quad (<280\text{—OK})$$

$$\text{Shop } Y: \quad (29 \quad 37 \quad 0)\begin{pmatrix}6\\5\\2\end{pmatrix} = 174 + 185 + 0 = 359 \quad (<360\text{—OK})$$

$$P = (29 \quad 37 \quad 0)\begin{pmatrix}15\\12\\9\end{pmatrix} = 435 + 444 + 0 = \$879.00$$

This profit figure is only \$6.00 less than our previous high of \$885.00—for A's only. Plenty of slack time in shop X; nearly none in shop Y.

2.
$$\text{Shop } X: \quad (36 \quad 24 \quad 0)\begin{pmatrix}4\\3\\4\end{pmatrix} = 204 + 72 + 0 = 276 \quad (<280\text{—OK})$$

$$\text{Shop } Y: \quad (36 \quad 24 \quad 0)\begin{pmatrix}6\\5\\2\end{pmatrix} = 216 + 120 + 0 = 336 \quad (<360\text{—OK})$$

$$P = (36 \quad 24 \quad 0)\begin{pmatrix}15\\12\\9\end{pmatrix} = 540 + 288 + 0 = \$828.00$$

Profit decreased from first try; therefore it would seem to be more advantageous to stress B and add some C's on the next try.

3.
$$\text{Shop } X: \quad (27 \quad 34 \quad 12)\begin{pmatrix}4\\3\\4\end{pmatrix} = 108 + 102 + 48 = 258 \quad (<280\text{—OK})$$

$$\text{Shop } Y: \quad (27 \quad 34 \quad 12)\begin{pmatrix}6\\5\\2\end{pmatrix} = 162 + 170 + 24 = 356 \quad (<360\text{—OK})$$

$$P = (27 \quad 34 \quad 12)\begin{pmatrix}15\\12\\9\end{pmatrix} = 405 + 408 + 108 = \$921.00$$

Now we seem to be moving in the right direction.

4.

$$\text{Shop } X: \quad (24 \quad 36 \quad 18)\begin{pmatrix}4\\3\\4\end{pmatrix} = 96 + 108 + 72 = 276 \quad (<280-\text{OK})$$

$$\text{Shop } Y: \quad (24 \quad 36 \quad 18)\begin{pmatrix}6\\5\\2\end{pmatrix} = 144 + 180 + 36 = 360 \quad (=360-\text{OK})$$

$$P = (24 \quad 36 \quad 18)\begin{pmatrix}15\\12\\9\end{pmatrix} = 360 + 432 + 162 = \$954.00$$

Trying a little more or less in any direction either exceeds the limits or decreases profits.

Solution:
Production: 24A, 36B, 18C
Profit: $954.00
Slack time: shop X—4 hours
              shop Y—0 hours

REVIEW QUESTIONS

8.5. What would be the cost factor of producing 24A, 36B, and 18C in the previous problem?

8.6. What would be the total gross income from the radios produced?

8.7. How many total hours will it take to produce:
(a) 24 A's
(b) 36 B's
(c) 18 C's

8.8. What would be the total net profit on the production listed in problem 8.7?

MINIMIZING COSTS

The techniques used for minimizing costs of an operation are identical to those used for maximizing profit. The first step is to order all of the available facts; the second step is to determine the restrictions to be imposed on the problem; the third step is to set up the equation (or equations) required to solve the problem. If the facts and restrictions are thoroughly understood, the equations become self-evident in most cases.

Although the most common use of the technique of minimizing or maximizing is for cost or profit of a business, it does not necessarily have that purpose. The following example is a case in point.

EXAMPLE

Problem Statement

Each printed page in a book is to have a top margin of two inches, a bottom margin of two inches, and side margins of one inch each. If the printed area is restricted to 40 square inches and the total page width should be between 6 and $6\frac{1}{2}$ inches, what is the smallest size page that can be used?

Let

$$L = \text{length of printed material in inches}$$
$$W = \text{width of printed material in inches}$$
$$A = \text{area of page in square inches } (L \times W = A)$$
$$S = \text{size of page}$$

Then

$$L + 4 = \text{length of page in inches}$$
$$W + 2 = \text{width of page in inches}$$

Given

$$LW = 40$$
$$A = (L + 4)(W + 2)$$
$$W > 6, < 6\frac{1}{2}$$

Minimize A to get the smallest size page possible.

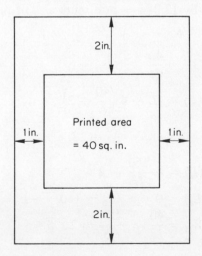

Estimation method:

1. The first idea that comes to mind is that $5 \times 8 = 40$

$$W = 5 + 2 = \quad 7 \qquad (W > 6\tfrac{1}{2}\text{—not OK})$$
$$L = 8 + 4 = 12$$
$$S = 7 \text{ by } 12$$

For the next try, lower width to 4.5 to meet the $6\tfrac{1}{2}$ requirement.

---

2. Restriction: $\qquad WL = 40$

$$(4.5)(8.9) = 39.95 \qquad (<40 \text{ by } .05)$$
$$W = 4.5 + 2 = \quad 6.5$$
$$L = 8.9 + 4 = 12.9$$
$$S = 6.5 \text{ by } 12.9$$

Increase the number of decimal places to get closer to 40.

---

3. $\qquad (4.03)(9.94) = 40.0582 \qquad (>40 \text{ by } .05)$

Increase decimal places once more.

---

4. $\qquad (4.024)(9.938) = 39.990 \dots (\text{close enough to } 40)$

$$W = 4.024 + 2 = \quad 6.024$$
$$L = 9.938 + 4 = 13.938$$

W is within limits, but maybe L can be made smaller without exceeding the W limit.

---

5. $\qquad (4.472)(8.944) = 39.997 \dots (\text{can't get closer to } 40)$

$$W = 4.472 + 2 = \quad 6.472$$
$$L = 8.944 + 4 = 12.944$$
$$S = 6.472 \text{ by } 12.944 = \text{minimum size of page}$$

Algebraic method:

$$\therefore \quad W = \frac{40}{L}$$

$$\therefore \quad A(L) = (L + 4)\left(\frac{40}{L} + 2\right)$$

$$= 48 + \frac{160}{L} + 2L$$

$$\longrightarrow A'(L) = -\frac{160}{L^2} + 2$$

$$A'(L_0) = 0 \Longrightarrow 2L_0^2 = 160 \Longrightarrow L_0 = 4\sqrt{5}$$

$$W_0 = \frac{40}{L_0} = \frac{10}{\sqrt{5}} = 2\sqrt{5}$$

$$A''(L) = +\frac{320}{L^3}$$

$$\longrightarrow A''(L_0) = \frac{320}{320\sqrt{5}}\frac{\sqrt{5}}{5} > 0$$

$\longrightarrow A(L_0)$ is minimum area

$$\sqrt{5} = 2.236$$

$$\longrightarrow \begin{cases} W_0 + 2 = & 6.472 \text{ inches} \\ L_0 + 4 = & 12.944 \text{ inches} \end{cases}$$

$$S = 6.472 \text{ by } 12.944$$

The important point to remember about linear programming problems is that the facts and restrictions specified for a particular problem must be properly analyzed and placed into workable equations.

The problem requirements are the same, regardless of the method of solution selected. The actual arithmetic is never done by the programmer anyway because the computer does a better job of calculating arithmetic results than any human being can ever do.

The job of the programmer is to set up each individual problem based on its own specific facts and restrictions. Once this is accomplished, he can "let the computer do it."

A Hexadecimal-Decimal Conversion Table

	0	1	2	3	4	5	6	7	8	9	A	B	C	D	E	F
00_	0000	0001	0002	0003	0004	0005	0006	0007	0008	0009	0010	0011	0012	0013	0014	0015
01_	0016	0017	0018	0019	0020	0021	0022	0023	0024	0025	0026	0027	0028	0029	0030	0031
02_	0032	0033	0034	0035	0036	0037	0038	0039	0040	0041	0042	0043	0044	0045	0046	0047
03_	0048	0049	0050	0051	0052	0053	0054	0055	0056	0057	0058	0059	0060	0061	0062	0063
04_	0064	0065	0066	0067	0068	0069	0070	0071	0072	0073	0074	0075	0076	0077	0078	0079
05_	0080	0081	0082	0083	0084	0085	0086	0087	0088	0089	0090	0091	0092	0093	0094	0095
06_	0096	0097	0098	0099	0100	0101	0102	0103	0104	0105	0106	0107	0108	0109	0110	0111
07_	0112	0113	0114	0115	0116	0117	0118	0119	0120	0121	0122	0123	0124	0125	0126	0127
08_	0128	0129	0130	0131	0132	0133	0134	0135	0136	0137	0138	0139	0140	0141	0142	0143
09_	0144	0145	0146	0147	0148	0149	0150	0151	0152	0153	0154	0155	0156	0157	0158	0159
0A_	0160	0161	0162	0163	0164	0165	0166	0167	0168	0169	0170	0171	0172	0173	0174	0175
0B_	0176	0177	0178	0179	0180	0181	0182	0183	0184	0185	0186	0187	0188	0189	0190	0191
0C_	0192	0193	0194	0195	0196	0197	0198	0199	0200	0201	0202	0203	0204	0205	0206	0207
0D_	0208	0209	0210	0211	0212	0213	0214	0215	0216	0217	0218	0219	0220	0221	0222	0223
0E_	0224	0225	0226	0227	0228	0229	0230	0231	0232	0233	0234	0235	0236	0237	0238	0239
0F_	0240	0241	0242	0243	0244	0245	0246	0247	0248	0249	0250	0251	0252	0253	0254	0255
10_	0256	0257	0258	0259	0260	0261	0262	0263	0264	0265	0266	0267	0268	0269	0270	0271
11_	0272	0273	0274	0275	0276	0277	0278	0279	0280	0281	0282	0283	0284	0285	0286	0287
12_	0288	0289	0290	0291	0292	0293	0294	0295	0296	0297	0298	0299	0300	0301	0302	0303
13_	0304	0305	0306	0307	0308	0309	0310	0311	0312	0313	0314	0315	0316	0317	0318	0319
14_	0320	0321	0322	0323	0324	0325	0326	0327	0328	0329	0330	0331	0332	0333	0334	0335
15_	0336	0337	0338	0339	0340	0341	0342	0343	0344	0345	0346	0347	0348	0349	0350	0351
16_	0352	0353	0354	0355	0356	0357	0358	0359	0360	0361	0362	0363	0364	0365	0366	0367
17_	0368	0369	0370	0371	0372	0373	0374	0375	0376	0377	0378	0379	0380	0381	0382	0383
18_	0384	0385	0386	0387	0388	0389	0390	0391	0392	0393	0394	0395	0396	0397	0398	0399
19_	0400	0401	0402	0403	0404	0405	0406	0407	0408	0409	0410	0411	0412	0413	0414	0415
1A_	0416	0417	0418	0419	0420	0421	0422	0423	0424	0425	0426	0427	0428	0429	0430	0431
1B_	0432	0433	0434	0435	0436	0437	0438	0439	0440	0441	0442	0443	0444	0445	0446	0447
1C_	0448	0449	0450	0451	0452	0453	0454	0455	0456	0457	0458	0459	0460	0461	0462	0463
1D_	0464	0465	0466	0467	0468	0469	0470	0471	0472	0473	0474	0475	0476	0477	0478	0479
1E_	0480	0481	0482	0483	0484	0485	0486	0487	0488	0489	0490	0491	0492	0493	0494	0495
1F_	0496	0497	0498	0499	0500	0501	0502	0503	0504	0505	0506	0507	0508	0509	0510	0511

	0	1	2	3	4	5	6	7	8	9	A	B	C	D	E	F
20_	0512	0513	0514	0515	0516	0517	0518	0519	0520	0521	0522	0523	0524	0525	0526	0527
21_	0528	0529	0530	0531	0532	0533	0534	0535	0536	0537	0538	0539	0540	0541	0542	0543
22_	0544	0545	0546	0547	0548	0549	0550	0551	0552	0553	0554	0555	0556	0557	0558	0559
23_	0560	0561	0562	0563	0564	0565	0566	0567	0568	0569	0570	0571	0572	0573	0574	0575
24_	0576	0577	0578	0579	0580	0581	0582	0583	0584	0585	0586	0587	0588	0589	0590	0591
25_	0592	0593	0594	0595	0596	0597	0598	0599	0600	0601	0602	0603	0604	0605	0606	0607
26_	0608	0609	0610	0611	0612	0613	0614	0615	0616	0617	0618	0619	0620	0621	0622	0623
27_	0624	0625	0626	0627	0628	0629	0630	0631	0632	0633	0634	0635	0636	0637	0638	0639
28_	0640	0641	0642	0643	0644	0645	0646	0647	0648	0649	0650	0651	0652	0653	0654	0655
29_	0656	0657	0658	0659	0660	0661	0662	0663	0664	0665	0666	0667	0668	0669	0670	0671
2A_	0672	0673	0674	0675	0676	0677	0678	0679	0680	0681	0682	0683	0684	0685	0686	0687
2B_	0688	0689	0690	0691	0692	0693	0694	0695	0696	0697	0698	0699	0700	0701	0702	0703
2C_	0704	0705	0706	0707	0708	0709	0710	0711	0712	0713	0714	0715	0716	0717	0718	0719
2D_	0720	0721	0722	0723	0724	0725	0726	0727	0728	0729	0730	0731	0732	0733	0734	0735
2E_	0736	0737	0738	0739	0740	0741	0742	0743	0744	0745	0746	0747	0748	0749	0750	0751
2F_	0752	0753	0754	0755	0756	0757	0758	0759	0760	0761	0762	0763	0764	0765	0766	0767
30_	0768	0769	0770	0771	0772	0773	0774	0775	0776	0777	0778	0779	0780	0781	0782	0783
31_	0784	0785	0786	0787	0788	0789	0790	0791	0792	0793	0794	0795	0796	0797	0798	0799
32_	0800	0801	0802	0803	0804	0805	0806	0807	0808	0809	0810	0811	0812	0813	0814	0815
33_	0816	0817	0818	0819	0820	0821	0822	0823	0824	0825	0826	0827	0828	0829	0830	0831
34_	0832	0833	0834	0835	0836	0837	0838	0839	0840	0841	0842	0843	0844	0845	0846	0847
35_	0848	0849	0850	0851	0852	0853	0854	0855	0856	0857	0858	0859	0860	0861	0862	0863
36_	0864	0865	0866	0867	0868	0869	0870	0871	0872	0873	0874	0875	0876	0877	0878	0879
37_	0880	0881	0882	0883	0884	0885	0886	0887	0888	0889	0890	0891	0892	0893	0894	0895
38_	0896	0897	0898	0899	0900	0901	0902	0903	0904	0905	0906	0907	0908	0909	0910	0911
39_	0912	0913	0914	0915	0916	0917	0918	0919	0920	0921	0922	0923	0924	0925	0926	0927
3A_	0928	0929	0930	0931	0932	0933	0934	0935	0936	0937	0938	0939	0940	0941	0942	0943
3B_	0944	0945	0946	0947	0948	0949	0950	0951	0952	0953	0954	0955	0956	0957	0958	0959
3C_	0960	0961	0962	0963	0964	0965	0966	0967	0968	0969	0970	0971	0972	0973	0974	0975
3D_	0976	0977	0978	0979	0980	0981	0982	0983	0984	0985	0986	0987	0988	0989	0990	0991
3E_	0992	0993	0994	0995	0996	0997	0998	0999	1000	1001	1002	1003	1004	1005	1006	1007
3F_	1008	1009	1010	1011	1012	1013	1014	1015	1016	1017	1018	1019	1020	1021	1022	1023

	0	1	2	3	4	5	6	7	8	9	A	B	C	D	E	F
40_	1024	1025	1026	1027	1028	1029	1030	1031	1032	1033	1034	1035	1036	1037	1038	1039
41_	1040	1041	1042	1043	1044	1045	1046	1047	1048	1049	1050	1051	1052	1053	1054	1055
42_	1056	1057	1058	1059	1060	1061	1062	1063	1064	1065	1066	1067	1068	1069	1070	1071
43_	1072	1073	1074	1075	1076	1077	1078	1079	1080	1081	1082	1083	1084	1085	1086	1087
44_	1088	1089	1090	1091	1092	1093	1094	1095	1096	1097	1098	1099	1100	1101	1102	1103
45_	1104	1105	1106	1107	1108	1109	1110	1111	1112	1113	1114	1115	1116	1117	1118	1119
46_	1120	1121	1122	1123	1124	1125	1126	1127	1128	1129	1130	1131	1132	1133	1134	1135
47_	1136	1137	1138	1139	1140	1141	1142	1143	1144	1145	1146	1147	1148	1149	1150	1151
48_	1152	1153	1154	1155	1156	1157	1158	1159	1160	1161	1162	1163	1164	1165	1166	1167
49_	1168	1169	1170	1171	1172	1173	1174	1175	1176	1177	1178	1179	1180	1181	1182	1183
4A_	1184	1185	1186	1187	1188	1189	1190	1191	1192	1193	1194	1195	1196	1197	1198	1199
4B_	1200	1201	1202	1203	1204	1205	1206	1207	1208	1209	1210	1211	1212	1213	1214	1215
4C_	1216	1217	1218	1219	1220	1221	1222	1223	1224	1225	1226	1227	1228	1229	1230	1231
4D_	1232	1233	1234	1235	1236	1237	1238	1239	1240	1241	1242	1243	1244	1245	1246	1247
4E_	1248	1249	1250	1251	1252	1253	1254	1255	1256	1257	1258	1259	1260	1261	1262	1263
4F_	1264	1265	1266	1267	1268	1269	1270	1271	1272	1273	1274	1275	1276	1277	1278	1279
50_	1280	1281	1282	1283	1284	1285	1286	1287	1288	1289	1290	1291	1292	1293	1294	1295
51_	1296	1297	1298	1299	1300	1301	1302	1303	1304	1305	1306	1307	1308	1309	1310	1311
52_	1312	1313	1314	1315	1316	1317	1318	1319	1320	1321	1322	1323	1324	1325	1326	1327
53_	1328	1329	1330	1331	1332	1333	1334	1335	1336	1337	1338	1339	1340	1341	1342	1343
54_	1344	1345	1346	1347	1348	1349	1350	1351	1352	1353	1354	1355	1356	1357	1358	1359
55_	1360	1361	1362	1363	1364	1365	1366	1367	1368	1369	1370	1371	1372	1373	1374	1375
56_	1376	1377	1378	1379	1380	1381	1382	1383	1384	1385	1386	1387	1388	1389	1390	1391
57_	1392	1393	1394	1395	1396	1397	1398	1399	1400	1401	1402	1403	1404	1405	1406	1407
58_	1408	1409	1410	1411	1412	1413	1414	1415	1416	1417	1418	1419	1420	1421	1422	1423
59_	1424	1425	1426	1427	1428	1429	1430	1431	1432	1433	1434	1435	1436	1437	1438	1439
5A_	1440	1441	1442	1443	1444	1445	1446	1447	1448	1449	1450	1451	1452	1453	1454	1455
5B_	1456	1457	1458	1459	1460	1461	1462	1463	1464	1465	1466	1467	1468	1469	1470	1471
5C_	1472	1473	1474	1475	1476	1477	1478	1479	1480	1481	1482	1483	1484	1485	1486	1487
5D_	1488	1489	1490	1491	1492	1493	1494	1495	1496	1497	1498	1499	1500	1501	1502	1503
5E_	1504	1505	1506	1507	1508	1509	1510	1511	1512	1513	1514	1515	1516	1517	1518	1519
5F_	1520	1521	1522	1523	1524	1525	1526	1527	1528	1529	1530	1531	1532	1533	1534	1535

	0	1	2	3	4	5	6	7	8	9	A	B	C	D	E	F
60_	1536	1537	1538	1539	1540	1541	1542	1543	1544	1545	1546	1547	1548	1549	1550	1551
61_	1552	1553	1554	1555	1556	1557	1558	1559	1560	1561	1562	1563	1564	1565	1566	1567
62_	1568	1569	1570	1571	1572	1573	1574	1575	1576	1577	1578	1579	1580	1581	1582	1583
63_	1584	1585	1586	1587	1588	1589	1590	1591	1592	1593	1594	1595	1596	1597	1598	1599
64_	1600	1601	1602	1603	1604	1605	1606	1607	1608	1609	1610	1611	1612	1613	1614	1615
65_	1616	1617	1618	1619	1620	1621	1622	1623	1624	1625	1626	1627	1628	1629	1630	1631
66_	1632	1633	1634	1635	1636	1637	1638	1639	1640	1641	1642	1643	1644	1645	1646	1647
67_	1648	1649	1650	1651	1652	1653	1654	1655	1656	1657	1658	1659	1660	1661	1662	1663
68_	1664	1665	1666	1667	1668	1669	1670	1671	1672	1673	1674	1675	1676	1677	1678	1679
69_	1680	1681	1682	1683	1684	1685	1686	1687	1688	1689	1690	1691	1692	1693	1694	1695
6A_	1696	1697	1698	1699	1700	1701	1702	1703	1704	1705	1706	1707	1708	1709	1710	1711
6B_	1712	1713	1714	1715	1716	1717	1718	1719	1720	1721	1722	1723	1724	1725	1726	1727
6C_	1728	1729	1730	1731	1732	1733	1734	1735	1736	1737	1738	1739	1740	1741	1742	1743
6D_	1744	1745	1746	1747	1748	1749	1750	1751	1752	1753	1754	1755	1756	1757	1758	1759
6E_	1760	1761	1762	1763	1764	1765	1766	1767	1768	1769	1770	1771	1772	1773	1774	1775
6F_	1776	1777	1778	1779	1780	1781	1782	1783	1784	1785	1786	1787	1788	1789	1790	1791
70_	1792	1793	1794	1795	1796	1797	1798	1799	1800	1801	1802	1803	1804	1805	1806	1807
71_	1808	1809	1810	1811	1812	1813	1814	1815	1816	1817	1818	1819	1820	1821	1822	1823
72_	1824	1825	1826	1827	1828	1829	1830	1831	1832	1833	1834	1835	1836	1837	1838	1839
73_	1840	1841	1842	1843	1844	1845	1846	1847	1848	1849	1850	1851	1852	1853	1854	1855
74_	1856	1857	1858	1859	1860	1861	1862	1863	1864	1865	1866	1867	1868	1869	1870	1871
75_	1872	1873	1874	1875	1876	1877	1878	1879	1880	1881	1882	1883	1884	1885	1886	1887
76_	1888	1889	1890	1891	1892	1893	1894	1895	1896	1897	1898	1899	1900	1901	1902	1903
77_	1904	1905	1906	1907	1908	1909	1910	1911	1912	1913	1914	1915	1916	1917	1918	1919
78_	1920	1921	1922	1923	1924	1925	1926	1927	1928	1929	1930	1931	1932	1933	1934	1935
79_	1936	1937	1938	1939	1940	1941	1942	1943	1944	1945	1946	1947	1948	1949	1950	1951
7A_	1952	1953	1954	1955	1956	1957	1958	1959	1960	1961	1962	1963	1964	1965	1966	1967
7B_	1968	1969	1970	1971	1972	1973	1974	1975	1976	1977	1978	1979	1980	1981	1982	1983
7C_	1984	1985	1986	1987	1988	1989	1990	1991	1992	1993	1994	1995	1996	1997	1998	1999
7D_	2000	2001	2002	2003	2004	2005	2006	2007	2008	2009	2010	2011	2012	2013	2014	2015
7E_	2016	2017	2018	2019	2020	2021	2022	2023	2024	2025	2026	2027	2028	2029	2030	2031
7F_	2032	2033	2034	2035	2036	2037	2038	2039	2040	2041	2042	2043	2044	2045	2046	2047

	0	1	2	3	4	5	6	7	8	9	A	B	C	D	E	F
80_	2048	2049	2050	2051	2052	2053	2054	2055	2056	2057	2058	2059	2060	2061	2062	2063
81_	2064	2065	2066	2067	2068	2069	2070	2071	2072	2073	2074	2075	2076	2077	2078	2079
82_	2080	2081	2082	2083	2084	2085	2086	2087	2088	2089	2090	2091	2092	2093	2094	2095
83_	2096	2097	2098	2099	2100	2101	2102	2103	2104	2105	2106	2107	2108	2109	2110	2111
84_	2112	2113	2114	2115	2116	2117	2118	2119	2120	2121	2122	2123	2124	2125	2126	2127
85_	2128	2129	2130	2131	2132	2133	2134	2135	2136	2137	2138	2139	2140	2141	2142	2143
86_	2144	2145	2146	2147	2148	2149	2150	2151	2152	2153	2154	2155	2156	2157	2158	2159
87_	2160	2161	2162	2163	2164	2165	2166	2167	2168	2169	2170	2171	2172	2173	2174	2175
88_	2176	2177	2178	2179	2180	2181	2182	2183	2184	2185	2186	2187	2188	2189	2190	2191
89_	2192	2193	2194	2195	2196	2197	2198	2199	2200	2201	2202	2203	2204	2205	2206	2207
8A_	2208	2209	2210	2211	2212	2213	2214	2215	2216	2217	2218	2219	2220	2221	2222	2223
8B_	2224	2225	2226	2227	2228	2229	2230	2231	2232	2233	2234	2235	2236	2237	2238	2239
8C_	2240	2241	2242	2243	2244	2245	2246	2247	2248	2249	2250	2251	2252	2253	2254	2255
8D_	2256	2257	2258	2259	2260	2261	2262	2263	2264	2265	2266	2267	2268	2269	2270	2271
8E_	2272	2273	2274	2275	2276	2277	2278	2279	2280	2281	2282	2283	2284	2285	2286	2287
8F_	2288	2289	2290	2291	2292	2293	2294	2295	2296	2297	2298	2299	2300	2301	2302	2303
90_	2304	2305	2306	2307	2308	2309	2310	2311	2312	2313	2314	2315	2316	2317	2318	2319
91_	2320	2321	2322	2323	2324	2325	2326	2327	2328	2329	2330	2331	2332	2333	2334	2335
92_	2336	2337	2338	2339	2340	2341	2342	2343	2344	2345	2346	2347	2348	2349	2350	2351
93_	2352	2353	2354	2355	2356	2357	2358	2359	2360	2361	2362	2363	2364	2365	2366	2367
94_	2368	2369	2370	2371	2372	2373	2374	2375	2376	2377	2378	2379	2380	2381	2382	2383
95_	2384	2385	2386	2387	2388	2389	2390	2391	2392	2393	2394	2395	2396	2397	2398	2399
96_	2400	2401	2402	2403	2404	2405	2406	2407	2408	2409	2410	2411	2412	2413	2414	2415
97_	2416	2417	2418	2419	2420	2421	2422	2423	2424	2425	2426	2427	2428	2429	2430	2431
98_	2432	2433	2434	2435	2436	2437	2438	2439	2440	2441	2442	2443	2444	2445	2446	2447
99_	2448	2449	2450	2451	2452	2453	2454	2455	2456	2457	2458	2459	2460	2461	2462	2463
9A_	2464	2465	2466	2467	2468	2469	2470	2471	2472	2473	2474	2475	2476	2477	2478	2479
9B_	2480	2481	2482	2483	2484	2485	2486	2487	2488	2489	2490	2491	2492	2493	2494	2495
9C_	2496	2497	2498	2499	2500	2501	2502	2503	2504	2505	2506	2507	2508	2509	2510	2511
9D_	2512	2513	2514	2515	2516	2517	2518	2519	2520	2521	2522	2523	2524	2525	2526	2527
9E_	2528	2529	2530	2531	2532	2533	2534	2535	2536	2537	2538	2539	2540	2541	2542	2543
9F_	2544	2545	2546	2547	2548	2549	2550	2551	2552	2553	2554	2555	2556	2557	2558	2559

	0	1	2	3	4	5	6	7	8	9	A	B	C	D	E	F
A0_	2560	2561	2562	2563	2564	2565	2566	2567	2568	2569	2570	2571	2572	2573	2574	2575
A1_	2576	2577	2578	2579	2580	2581	2582	2583	2584	2585	2586	2587	2588	2589	2590	2591
A2_	2592	2593	2594	2595	2596	2597	2598	2599	2600	2601	2602	2603	2604	2605	2606	2607
A3_	2608	2609	2610	2611	2612	2613	2614	2615	2616	2617	2618	2619	2620	2621	2622	2623
A4_	2624	2625	2626	2627	2628	2629	2630	2631	2632	2633	2634	2635	2636	2637	2638	2639
A5_	2640	2641	2642	2643	2644	2645	2646	2647	2648	2649	2650	2651	2652	2653	2654	2655
A6_	2656	2657	2658	2659	2660	2661	2662	2663	2664	2665	2666	2667	2668	2669	2670	2671
A7_	2672	2673	2674	2675	2676	2677	2678	2679	2680	2681	2682	2683	2684	2685	2686	2687
A8_	2688	2689	2690	2691	2692	2693	2694	2695	2696	2697	2698	2699	2700	2701	2702	2703
A9_	2704	2705	2706	2707	2708	2709	2710	2711	2712	2713	2714	2715	2716	2717	2718	2719
AA_	2720	2721	2722	2723	2724	2725	2726	2727	2728	2729	2730	2731	2732	2733	2734	2735
AB_	2736	2737	2738	2739	2740	2741	2742	2743	2744	2745	2746	2747	2748	2749	2750	2751
AC_	2752	2753	2754	2755	2756	2757	2758	2759	2760	2761	2762	2763	2764	2765	2766	2767
AD_	2768	2769	2770	2771	2772	2773	2774	2775	2776	2777	2778	2779	2780	2781	2782	2783
AE_	2784	2785	2786	2787	2788	2789	2790	2791	2792	2793	2794	2795	2796	2797	2798	2799
AF_	2800	2801	2802	2803	2804	2805	2806	2807	2808	2809	2810	2811	2812	2813	2814	2815
B0_	2816	2817	2818	2819	2820	2821	2822	2823	2824	2825	2826	2827	2828	2829	2830	2831
B1_	2832	2833	2834	2835	2836	2837	2838	2839	2840	2841	2842	2843	2844	2845	2846	2847
B2_	2848	2849	2850	2851	2852	2853	2854	2855	2856	2857	2858	2859	2860	2861	2862	2863
B3_	2864	2865	2866	2867	2868	2869	2870	2871	2872	2873	2874	2875	2876	2877	2878	2879
B4_	2880	2881	2882	2883	2884	2885	2886	2887	2888	2889	2890	2891	2892	2893	2894	2895
B5_	2896	2897	2898	2899	2900	2901	2902	2903	2904	2905	2906	2907	2908	2909	2910	2911
B6_	2912	2913	2914	2915	2916	2917	2918	2919	2920	2921	2922	2923	2924	2925	2926	2927
B7_	2928	2929	2930	2931	2932	2933	2934	2935	2936	2937	2938	2939	2940	2941	2942	2943
B8_	2944	2945	2946	2947	2948	2949	2950	2951	2952	2953	2954	2955	2956	2957	2958	2959
B9_	2960	2961	2962	2963	2964	2965	2966	2967	2968	2969	2970	2971	2972	2973	2974	2975
BA_	2976	2977	2978	2979	2980	2981	2982	2983	2984	2985	2986	2987	2988	2989	2990	2991
BB_	2992	2993	2994	2995	2996	2997	2998	2999	3000	3001	3002	3003	3004	3005	3006	3007
BC_	3008	3009	3010	3011	3012	3013	3014	3015	3016	3017	3018	3019	3020	3021	3022	3023
BD_	3024	3025	3026	3027	3028	3029	3030	3031	3032	3033	3034	3035	3036	3037	3038	3039
BE_	3040	3041	3042	3043	3044	3045	3046	3047	3048	3049	3050	3051	3052	3053	3054	3055
BF_	3056	3057	3058	3059	3060	3061	3062	3063	3064	3065	3066	3067	3068	3069	3070	3071

	0	1	2	3	4	5	6	7	8	9	A	B	C	D	E	F
C0_	3072	3073	3074	3075	3076	3077	3078	3079	3080	3081	3082	3083	3084	3085	3086	3087
C1_	3088	3089	3090	3091	3092	3093	3094	3095	3096	3097	3098	3099	3100	3101	3102	3103
C2_	3104	3105	3106	3107	3108	3109	3110	3111	3112	3113	3114	3115	3116	3117	3118	3119
C3_	3120	3121	3122	3123	3124	3125	3126	3127	3128	3129	3130	3131	3132	3133	3134	3135
C4_	3136	3137	3138	3139	3140	3141	3142	3143	3144	3145	3146	3147	3148	3149	3150	3151
C5_	3152	3153	3154	3155	3156	3157	3158	3159	3160	3161	3162	3163	3164	3165	3166	3167
C6_	3168	3169	3170	3171	3172	3173	3174	3175	3176	3177	3178	3179	3180	3181	3182	3183
C7_	3184	3185	3186	3187	3188	3189	3190	3191	3192	3193	3194	3195	3196	3197	3198	3199
C8_	3200	3201	3202	3203	3204	3205	3206	3207	3208	3209	3210	3211	3212	3213	3214	3215
C9_	3216	3217	3218	3219	3220	3221	3222	3223	3224	3225	3226	3227	3228	3229	3230	3231
CA_	3232	3233	3234	3235	3236	3237	3238	3239	3240	3241	3242	3243	3244	3245	3246	3247
CB_	3248	3249	3250	3251	3252	3253	3254	3255	3256	3257	3258	3259	3260	3261	3262	3263
CC_	3264	3265	3266	3267	3268	3269	3270	3271	3272	3273	3274	3275	3276	3277	3278	3279
CD_	3280	3281	3282	3283	3284	3285	3286	3287	3288	3289	3290	3291	3292	3293	3294	3295
CE_	3296	3297	3298	3299	3300	3301	3302	3303	3304	3305	3306	3307	3308	3309	3310	3311
CF_	3312	3313	3314	3315	3316	3317	3318	3319	3320	3321	3322	3323	3324	3325	3326	3327
D0_	3328	3329	3330	3331	3332	3333	3334	3335	3336	3337	3338	3339	3340	3341	3342	3343
D1_	3344	3345	3346	3347	3348	3349	3350	3351	3352	3353	3354	3355	3356	3357	3358	3359
D2_	3360	3361	3362	3363	3364	3365	3366	3367	3368	3369	3370	3371	3372	3373	3374	3375
D3_	3376	3377	3378	3379	3380	3381	3382	3383	3384	3385	3386	3387	3388	3389	3390	3391
D4_	3392	3393	3394	3395	3396	3397	3398	3399	3400	3401	3402	3403	3404	3405	3406	3407
D5_	3408	3409	3410	3411	3412	3413	3414	3415	3416	3417	3418	3419	3420	3421	3422	3423
D6_	3424	3425	3426	3427	3428	3429	3430	3431	3432	3433	3434	3435	3436	3437	3438	3439
D7_	3440	3441	3442	3443	3444	3445	3446	3447	3448	3449	3450	3451	3452	3453	3454	3455
D8_	3456	3457	3458	3459	3460	3461	3462	3463	3464	3465	3466	3467	3468	3469	3470	3471
D9_	3472	3473	3474	3475	3476	3477	3478	3479	3480	3481	3482	3483	3484	3485	3486	3487
DA_	3488	3489	3490	3491	3492	3493	3494	3495	3496	3497	3498	3499	3500	3501	3502	3503
DB_	3504	3505	3506	3507	3508	3509	3510	3511	3512	3513	3514	3515	3516	3517	3518	3519
DC_	3520	3521	3522	3523	3524	3525	3526	3527	3528	3529	3530	3531	3532	3533	3534	3535
DD_	3536	3537	3538	3539	3540	3541	3542	3543	3544	3545	3546	3547	3548	3549	3550	3551
DE_	3552	3553	3554	3555	3556	3557	3558	3559	3560	3561	3562	3563	3564	3565	3566	3567
DF_	3568	3569	3570	3571	3572	3573	3574	3575	3576	3577	3578	3579	3580	3581	3582	3583

	0	1	2	3	4	5	6	7	8	9	A	B	C	D	E	F
E0_	3584	3585	3586	3587	3588	3589	3590	3591	3592	3593	3594	3595	3596	3597	3598	3599
E1_	3600	3601	3602	3603	3604	3605	3606	3607	3608	3609	3610	3611	3612	3613	3614	3615
E2_	3616	3617	3618	3619	3620	3621	3622	3623	3624	3625	3626	3627	3628	3629	3630	3631
E3_	3632	3633	3634	3635	3636	3637	3638	3639	3640	3641	3642	3643	3644	3645	3646	3647
E4_	3648	3649	3650	3651	3652	3653	3654	3655	3656	3657	3658	3659	3660	3661	3662	3663
E5_	3664	3665	3666	3667	3668	3669	3670	3671	3672	3673	3674	3675	3676	3677	3678	3679
E6_	3680	3681	3682	3683	3684	3685	3686	3687	3688	3689	3690	3691	3692	3693	3694	3695
E7_	3696	3697	3698	3699	3700	3701	3702	3703	3704	3705	3706	3707	3708	3709	3710	3711
E8_	3712	3713	3714	3715	3716	3717	3718	3719	3720	3721	3722	3723	3724	3725	3726	3727
E9_	3728	3729	3730	3731	3732	3733	3734	3735	3736	3737	3738	3739	3740	3741	3742	3743
EA_	3744	3745	3746	3747	3748	3749	3750	3751	3752	3753	3754	3755	3756	3757	3758	3759
EB_	3760	3761	3762	3763	3764	3765	3766	3767	3768	3769	3770	3771	3772	3773	3774	3775
EC_	3776	3777	3778	3779	3780	3781	3782	3783	3784	3785	3786	3787	3788	3789	3790	3791
ED_	3792	3793	3794	3795	3796	3797	3798	3799	3800	3801	3802	3803	3804	3805	3806	3807
EE_	3808	3809	3810	3811	3812	3813	3814	3815	3816	3817	3818	3819	3820	3821	3822	3823
EF_	3824	3825	3826	3827	3828	3829	3830	3831	3832	3833	3834	3835	3836	3837	3838	3839
F0_	3840	3841	3842	3843	3844	3845	3846	3847	3848	3849	3850	3851	3852	3853	3854	3855
F1_	3856	3857	3858	3859	3860	3861	3862	3863	3864	3865	3866	3867	3868	3869	3870	3871
F2_	3872	3873	3874	3875	3876	3877	3878	3879	3880	3881	3882	3883	3884	3885	3886	3887
F3_	3888	3889	3890	3891	3892	3893	3894	3895	3896	3897	3898	3899	3900	3901	3902	3903
F4_	3904	3905	3906	3907	3908	3909	3910	3911	3912	3913	3914	3915	3916	3917	3918	3919
F5_	3920	3921	3922	3923	3924	3925	3926	3927	3928	3929	3930	3931	3932	3933	3934	3935
F6_	3936	3937	3938	3939	3940	3941	3942	3943	3944	3945	3946	3947	3948	3949	3950	3951
F7_	3952	3953	3954	3955	3956	3957	3958	3959	3960	3961	3962	3963	3964	3965	3966	3967
F8_	3968	3969	3970	3971	3972	3973	3974	3975	3976	3977	3978	3979	3980	3981	3982	3983
F9_	3984	3985	3986	3987	3988	3989	3990	3991	3992	3993	3994	3995	3996	3997	3998	3999
FA_	4000	4001	4002	4003	4004	4005	4006	4007	4008	4009	4010	4011	4012	4013	4014	4015
FB_	4016	4017	4018	4019	4020	4021	4022	4023	4024	4025	4026	4027	4028	4029	4030	4031
FC_	4032	4033	4034	4035	4036	4037	4038	4039	4040	4041	4042	4043	4044	4045	4046	4047
FD_	4048	4049	4050	4051	4052	4053	4054	4055	4056	4057	4058	4059	4060	4061	4062	4063
FE_	4064	4065	4066	4067	4068	4069	4070	4071	4072	4073	4074	4075	4076	4077	4078	4079
FF_	4080	4081	4082	4083	4084	4085	4086	4087	4088	4089	4090	4091	4092	4093	4094	4095

B Octal-Decimal Integer Conversion Table

OCTAL	0	1	2	3	4	5	6	7
0000	0000	0001	0002	0003	0004	0005	0006	0007
0010	0008	0009	0010	0011	0012	0013	0014	0015
0020	0016	0017	0018	0019	0020	0021	0022	0023
0030	0024	0025	0026	0027	0028	0029	0030	0031
0040	0032	0033	0034	0035	0036	0037	0038	0039
0050	0040	0041	0042	0043	0044	0045	0046	0047
0060	0048	0049	0050	0051	0052	0053	0054	0055
0070	0056	0057	0058	0059	0060	0061	0062	0063
0100	0064	0065	0066	0067	0068	0069	0070	0071
0110	0072	0073	0074	0075	0076	0077	0078	0079
0120	0080	0081	0082	0083	0084	0085	0086	0087
0130	0088	0089	0090	0091	0092	0093	0094	0095
0140	0096	0097	0098	0099	0100	0101	0102	0103
0150	0104	0105	0106	0107	0108	0109	0110	0111
0160	0112	0113	0114	0115	0116	0117	0118	0119
0170	0120	0121	0122	0123	0124	0125	0126	0127
0200	0128	0129	0130	0131	0132	0133	0134	0135
0210	0136	0137	0138	0139	0140	0141	0142	0143
0220	0144	0145	0146	0147	0148	0149	0150	0151
0230	0152	0153	0154	0155	0156	0157	0158	0159
0240	0160	0161	0162	0163	0164	0165	0166	0167
0250	0168	0169	0170	0171	0172	0173	0174	0175
0260	0176	0177	0178	0179	0180	0181	0182	0183
0270	0184	0185	0186	0187	0188	0189	0190	0191
0300	0192	0193	0194	0195	0196	0197	0198	0199
0310	0200	0201	0202	0203	0204	0205	0206	0207
0320	0208	0209	0210	0211	0212	0213	0214	0215
0330	0216	0217	0218	0219	0220	0221	0222	0223
0340	0224	0225	0226	0227	0228	0229	0230	0231
0350	0232	0233	0234	0235	0236	0237	0238	0239
0360	0240	0241	0242	0243	0244	0245	0246	0247
0370	0248	0249	0250	0251	0252	0253	0254	0255

OCTAL	0	1	2	3	4	5	6	7
0400	0256	0257	0258	0259	0260	0261	0262	0263
0410	0264	0265	0266	0267	0268	0269	0270	0271
0420	0272	0273	0274	0275	0276	0277	0278	0279
0430	0280	0281	0282	0283	0284	0285	0286	0287
0440	0288	0289	0290	0291	0292	0293	0294	0295
0450	0296	0297	0298	0299	0300	0301	0302	0303
0460	0304	0305	0306	0307	0308	0309	0310	0311
0470	0312	0313	0314	0315	0316	0317	0318	0319
0500	0320	0321	0322	0323	0324	0325	0326	0327
0510	0328	0329	0330	0331	0332	0333	0334	0335
0520	0336	0337	0338	0339	0340	0341	0342	0343
0530	0344	0345	0346	0347	0348	0349	0350	0351
0540	0352	0353	0354	0355	0356	0357	0358	0359
0550	0360	0361	0362	0363	0364	0365	0366	0367
0560	0368	0369	0370	0371	0372	0373	0374	0375
0570	0376	0377	0378	0379	0380	0381	0382	0383
0600	0384	0385	0386	0387	0388	0389	0390	0391
0610	0392	0393	0394	0395	0396	0397	0398	0399
0620	0400	0401	0402	0403	0404	0405	0406	0407
0630	0408	0409	0410	0411	0412	0413	0414	0415
0640	0416	0417	0418	0419	0420	0421	0422	0423
0650	0424	0425	0426	0427	0428	0429	0430	0431
0660	0432	0433	0434	0435	0436	0437	0438	0439
0670	0440	0441	0442	0443	0444	0445	0446	0447
0700	0448	0449	0450	0451	0452	0453	0454	0455
0710	0456	0457	0458	0459	0460	0461	0462	0463
0720	0464	0465	0466	0467	0468	0469	0470	0471
0730	0472	0473	0474	0475	0476	0477	0478	0479
0740	0480	0481	0482	0483	0484	0485	0486	0487
0750	0488	0489	0490	0491	0492	0493	0494	0495
0760	0496	0497	0498	0499	0500	0501	0502	0503
0770	0504	0505	0506	0507	0508	0509	0510	0511

OCTAL	0	1	2	3	4	5	6	7
1000	0512	0513	0514	0515	0516	0517	0518	0519
1010	0520	0521	0522	0523	0524	0525	0526	0527
1020	0528	0529	0530	0531	0532	0533	0534	0535
1030	0536	0537	0538	0539	0540	0541	0542	0543
1040	0544	0545	0546	0547	0548	0549	0550	0551
1050	0552	0553	0554	0555	0556	0557	0558	0559
1060	0560	0561	0562	0563	0564	0565	0566	0567
1070	0568	0569	0570	0571	0572	0573	0574	0575
1100	0576	0577	0578	0579	0580	0581	0582	0583
1110	0584	0585	0586	0587	0588	0589	0590	0591
1120	0592	0593	0594	0595	0596	0597	0598	0599
1130	0600	0601	0602	0603	0604	0605	0606	0607
1140	0608	0609	0610	0611	0612	0613	0614	0615
1150	0616	0617	0618	0619	0620	0621	0622	0623
1160	0624	0625	0626	0627	0628	0629	0630	0631
1170	0632	0633	0634	0635	0636	0637	0638	0639
1200	0640	0641	0642	0643	0644	0645	0646	0647
1210	0648	0649	0650	0651	0652	0653	0654	0655
1220	0656	0657	0658	0659	0660	0661	0662	0663
1230	0664	0665	0666	0667	0668	0669	0670	0671
1240	0672	0673	0674	0675	0676	0677	0678	0679
1250	0680	0681	0682	0683	0684	0685	0686	0687
1260	0688	0689	0690	0691	0692	0693	0694	0695
1270	0696	0697	0698	0699	0700	0701	0702	0703
1300	0704	0705	0706	0707	0708	0709	0710	0711
1310	0712	0713	0714	0715	0716	0717	0718	0719
1320	0720	0721	0722	0723	0724	0725	0726	0727
1330	0728	0729	0730	0731	0732	0733	0734	0735
1340	0736	0737	0738	0739	0740	0741	0742	0743
1350	0744	0745	0746	0747	0748	0749	0750	0751
1360	0752	0753	0754	0755	0756	0757	0758	0759
1370	0760	0761	0762	0763	0764	0765	0766	0767

OCTAL	0	1	2	3	4	5	6	7
1400	0768	0769	0770	0771	0772	0773	0774	0775
1410	0776	0777	0778	0779	0780	0781	0782	0783
1420	0784	0785	0786	0787	0788	0789	0790	0791
1430	0792	0793	0794	0795	0796	0797	0798	0799
1440	0800	0801	0802	0803	0804	0805	0806	0807
1450	0808	0809	0810	0811	0812	0813	0814	0815
1460	0816	0817	0818	0819	0820	0821	0822	0823
1470	0824	0825	0826	0827	0828	0829	0830	0831
1500	0832	0833	0834	0835	0836	0837	0838	0839
1510	0840	0841	0842	0843	0844	0845	0846	0847
1520	0848	0849	0850	0851	0852	0853	0854	0855
1530	0856	0857	0858	0859	0860	0861	0862	0863
1540	0864	0865	0866	0867	0868	0869	0870	0871
1550	0872	0873	0874	0875	0876	0877	0878	0879
1560	0880	0881	0882	0883	0884	0885	0886	0887
1570	0888	0889	0890	0891	0892	0893	0894	0895
1600	0896	0897	0898	0899	0900	0901	0902	0903
1610	0904	0905	0906	0907	0908	0909	0910	0911
1620	0912	0913	0914	0915	0916	0917	0918	0919
1630	0920	0921	0922	0923	0924	0925	0926	0927
1640	0928	0929	0930	0931	0932	0933	0934	0935
1650	0936	0937	0938	0939	0940	0941	0942	0943
1660	0944	0945	0946	0947	0948	0949	0950	0951
1670	0952	0953	0954	0955	0956	0957	0958	0959
1700	0960	0961	0962	0963	0964	0965	0966	0967
1710	0968	0969	0970	0971	0972	0973	0974	0975
1720	0976	0977	0978	0979	0980	0981	0982	0983
1730	0984	0985	0986	0987	0988	0989	0990	0991
1740	0992	0993	0994	0995	0996	0997	0998	0999
1750	1000	1001	1002	1003	1004	1005	1006	1007
1760	1008	1009	1010	1011	1012	1013	1014	1015
1770	1016	1017	1018	1019	1020	1021	1022	1023

OCTAL	0	1	2	3	4	5	6	7
2000	1024	1025	1026	1027	1028	1029	1030	1031
2010	1032	1033	1034	1035	1036	1037	1038	1039
2020	1040	1041	1042	1043	1044	1045	1046	1047
2030	1048	1049	1050	1051	1052	1053	1054	1055
2040	1056	1057	1058	1059	1060	1061	1062	1063
2050	1064	1065	1066	1067	1068	1069	1070	1071
2060	1072	1073	1074	1075	1076	1077	1078	1079
2070	1080	1081	1082	1083	1084	1085	1086	1087
2100	1088	1089	1090	1091	1092	1093	1094	1095
2110	1096	1097	1098	1099	1100	1101	1102	1103
2120	1104	1105	1106	1107	1108	1109	1110	1111
2130	1112	1113	1114	1115	1116	1117	1118	1119
2140	1120	1121	1122	1123	1124	1125	1126	1127
2150	1128	1129	1130	1131	1132	1133	1134	1135
2160	1136	1137	1138	1139	1140	1141	1142	1143
2170	1144	1145	1146	1147	1148	1149	1150	1151
2200	1152	1153	1154	1155	1156	1157	1158	1159
2210	1160	1161	1162	1163	1164	1165	1166	1167
2220	1168	1169	1170	1171	1172	1173	1174	1175
2230	1176	1177	1178	1179	1180	1181	1182	1183
2240	1184	1185	1186	1187	1188	1189	1190	1191
2250	1192	1193	1194	1195	1196	1197	1198	1199
2260	1200	1201	1202	1203	1204	1205	1206	1207
2270	1208	1209	1210	1211	1212	1213	1214	1215
2300	1216	1217	1218	1219	1220	1221	1222	1223
2310	1224	1225	1226	1227	1228	1229	1230	1231
2320	1232	1233	1234	1235	1236	1237	1238	1239
2330	1240	1241	1242	1243	1244	1245	1246	1247
2340	1248	1249	1250	1251	1252	1253	1254	1255
2350	1256	1257	1258	1259	1260	1261	1262	1263
2360	1264	1265	1266	1267	1268	1269	1270	1271
2370	1272	1273	1274	1275	1276	1277	1278	1279

OCTAL	0	1	2	3	4	5	6	7
2400	1280	1281	1282	1283	1284	1285	1286	1287
2410	1288	1289	1290	1291	1292	1293	1294	1295
2420	1296	1297	1298	1299	1300	1301	1302	1303
2430	1304	1305	1306	1307	1308	1309	1310	1311
2440	1312	1313	1314	1315	1316	1317	1318	1319
2450	1320	1321	1322	1323	1324	1325	1326	1327
2460	1328	1329	1330	1331	1332	1333	1334	1335
2470	1336	1337	1338	1339	1340	1341	1342	1343
2500	1344	1345	1346	1347	1348	1349	1350	1351
2510	1352	1353	1354	1355	1356	1357	1358	1359
2520	1360	1361	1362	1363	1364	1365	1366	1367
2530	1368	1369	1370	1371	1372	1373	1374	1375
2540	1376	1377	1378	1379	1380	1381	1382	1383
2550	1384	1385	1386	1387	1388	1389	1390	1391
2560	1392	1393	1394	1395	1396	1397	1398	1399
2570	1400	1401	1402	1403	1404	1405	1406	1407
2600	1408	1409	1410	1411	1412	1413	1414	1415
2610	1416	1417	1418	1419	1420	1421	1422	1423
2620	1424	1425	1426	1427	1428	1429	1430	1431
2630	1432	1433	1434	1435	1436	1437	1438	1439
2640	1440	1441	1442	1443	1444	1445	1446	1447
2650	1448	1449	1450	1451	1452	1453	1454	1455
2660	1456	1457	1458	1459	1460	1461	1462	1463
2670	1464	1465	1466	1467	1468	1469	1470	1471
2700	1472	1473	1474	1475	1476	1477	1478	1479
2710	1480	1481	1482	1483	1484	1485	1486	1487
2720	1488	1489	1490	1491	1492	1493	1494	1495
2730	1496	1497	1498	1499	1500	1501	1502	1503
2740	1504	1505	1506	1507	1508	1509	1510	1511
2750	1512	1513	1514	1515	1516	1517	1518	1519
2760	1520	1521	1522	1523	1524	1525	1526	1527
2770	1528	1529	1530	1531	1532	1533	1534	1535

OCTAL	0	1	2	3	4	5	6	7
3000	1536	1537	1538	1539	1540	1541	1542	1543
3010	1544	1545	1546	1547	1548	1549	1550	1551
3020	1552	1553	1554	1555	1556	1557	1558	1559
3030	1560	1561	1562	1563	1564	1565	1566	1567
3040	1568	1569	1570	1571	1572	1573	1574	1575
3050	1576	1577	1578	1579	1580	1581	1582	1583
3060	1584	1585	1586	1587	1588	1589	1590	1591
3070	1592	1593	1594	1595	1596	1597	1598	1599
3100	1600	1601	1602	1603	1604	1605	1606	1607
3110	1608	1609	1610	1611	1612	1613	1614	1615
3120	1616	1617	1618	1619	1620	1621	1622	1623
3130	1624	1625	1626	1627	1628	1629	1630	1631
3140	1632	1633	1634	1635	1636	1637	1638	1639
3150	1640	1641	1642	1643	1644	1645	1646	1647
3160	1648	1649	1650	1651	1652	1653	1654	1655
3170	1656	1657	1658	1659	1660	1661	1662	1663
3200	1664	1665	1666	1667	1668	1669	1670	1671
3210	1672	1673	1674	1675	1676	1677	1678	1679
3220	1680	1681	1682	1683	1684	1685	1686	1687
3230	1688	1689	1690	1691	1692	1693	1694	1695
3240	1696	1697	1698	1699	1700	1701	1702	1703
3250	1704	1705	1706	1707	1708	1709	1710	1711
3260	1712	1713	1714	1715	1716	1717	1718	1719
3270	1720	1721	1722	1723	1724	1725	1726	1727
3300	1728	1729	1730	1731	1732	1733	1734	1735
3310	1736	1737	1738	1739	1740	1741	1742	1743
3320	1744	1745	1746	1747	1748	1749	1750	1751
3330	1752	1753	1754	1755	1756	1757	1758	1759
3340	1760	1761	1762	1763	1764	1765	1766	1767
3350	1768	1769	1770	1771	1772	1773	1774	1775
3360	1776	1777	1778	1779	1780	1781	1782	1783
3370	1784	1785	1786	1787	1788	1789	1790	1791

OCTAL	0	1	2	3	4	5	6	7
3400	1792	1793	1794	1795	1796	1797	1798	1799
3410	1800	1801	1802	1803	1804	1805	1806	1807
3420	1808	1809	1810	1811	1812	1813	1814	1815
3430	1816	1817	1818	1819	1820	1821	1822	1823
3440	1824	1825	1826	1827	1828	1829	1830	1831
3450	1832	1833	1834	1835	1836	1837	1838	1839
3460	1840	1841	1842	1843	1844	1845	1846	1847
3470	1848	1849	1850	1851	1852	1853	1854	1855
3500	1856	1857	1858	1859	1860	1861	1862	1863
3510	1864	1865	1866	1867	1868	1869	1870	1871
3520	1872	1873	1874	1875	1876	1877	1878	1879
3530	1880	1881	1882	1883	1884	1885	1886	1887
3540	1888	1889	1890	1891	1892	1893	1894	1895
3550	1896	1897	1898	1899	1900	1901	1902	1903
3560	1904	1905	1906	1907	1908	1909	1910	1911
3570	1912	1913	1914	1915	1916	1917	1918	1919
3600	1920	1921	1922	1923	1924	1925	1926	1927
3610	1928	1929	1930	1931	1932	1933	1934	1935
3620	1936	1937	1938	1939	1940	1941	1942	1943
3630	1944	1945	1946	1947	1948	1949	1950	1951
3640	1952	1953	1954	1955	1956	1957	1958	1959
3650	1960	1961	1962	1963	1964	1965	1966	1967
3660	1968	1969	1970	1971	1972	1973	1974	1975
3670	1976	1977	1978	1979	1980	1981	1982	1983
3700	1984	1985	1986	1987	1988	1989	1990	1991
3710	1992	1993	1994	1995	1996	1997	1998	1999
3720	2000	2001	2002	2003	2004	2005	2006	2007
3730	2008	2009	2010	2011	2012	2013	2014	2015
3740	2016	2017	2018	2019	2020	2021	2022	2023
3750	2024	2025	2026	2027	2028	2029	2030	2031
3760	2032	2033	2034	2035	2036	2037	2038	2039
3770	2040	2041	2042	2043	2044	2045	2046	2047

OCTAL	0	1	2	3	4	5	6	7
4000	2048	2049	2050	2051	2052	2053	2054	2055
4010	2056	2057	2058	2059	2060	2061	2062	2063
4020	2064	2065	2066	2067	2068	2069	2070	2071
4030	2072	2073	2074	2075	2076	2077	2078	2079
4040	2080	2081	2082	2083	2084	2085	2086	2087
4050	2088	2089	2090	2091	2092	2093	2094	2095
4060	2096	2097	2098	2099	2100	2101	2102	2103
4070	2104	2105	2106	2107	2108	2109	2110	2111
4100	2112	2113	2114	2115	2116	2117	2118	2119
4110	2120	2121	2122	2123	2124	2125	2126	2127
4120	2128	2129	2130	2131	2132	2133	2134	2135
4130	2136	2137	2138	2139	2140	2141	2142	2143
4140	2144	2145	2146	2147	2148	2149	2150	2151
4150	2152	2153	2154	2155	2156	2157	2158	2159
4160	2160	2161	2162	2163	2164	2165	2166	2167
4170	2168	2169	2170	2171	2172	2173	2174	2175
4200	2176	2177	2178	2179	2180	2181	2182	2183
4210	2184	2185	2186	2187	2188	2189	2190	2191
4220	2192	2193	2194	2195	2196	2197	2198	2199
4230	2200	2201	2202	2203	2204	2205	2206	2207
4240	2208	2209	2210	2211	2212	2213	2214	2215
4250	2216	2217	2218	2219	2220	2221	2222	2223
4260	2224	2225	2226	2227	2228	2229	2230	2231
4270	2232	2233	2234	2235	2236	2237	2238	2239
4300	2240	2241	2242	2243	2244	2245	2246	2247
4310	2248	2249	2250	2251	2252	2253	2254	2255
4320	2256	2257	2258	2259	2260	2261	2262	2263
4330	2264	2265	2266	2267	2268	2269	2270	2271
4340	2272	2273	2274	2275	2276	2277	2278	2279
4350	2280	2281	2282	2283	2284	2285	2286	2287
4360	2288	2289	2290	2291	2292	2293	2294	2295
4370	2296	2297	2298	2299	2300	2301	2302	2303

OCTAL	0	1	2	3	4	5	6	7
4400	2304	2305	2306	2307	2308	2309	2310	2311
4410	2312	2313	2314	2315	2316	2317	2318	2319
4420	2320	2321	2322	2323	2324	2325	2326	2327
4430	2328	2329	2330	2331	2332	2333	2334	2335
4440	2336	2337	2338	2339	2340	2341	2342	2343
4450	2344	2345	2346	2347	2348	2349	2350	2351
4460	2352	2353	2354	2355	2356	2357	2358	2359
4470	2360	2361	2362	2363	2364	2365	2366	2367
4500	2368	2369	2370	2371	2372	2373	2374	2375
4510	2376	2377	2378	2379	2380	2381	2382	2383
4520	2384	2385	2386	2387	2388	2389	2390	2391
4530	2392	2393	2394	2395	2396	2397	2398	2399
4540	2400	2401	2402	2403	2404	2405	2406	2407
4550	2408	2409	2410	2411	2412	2413	2414	2415
4560	2416	2417	2418	2419	2420	2421	2422	2423
4570	2424	2425	2426	2427	2428	2429	2430	2431
4600	2432	2433	2434	2435	2436	2437	2438	2439
4610	2440	2441	2442	2443	2444	2445	2446	2447
4620	2448	2449	2450	2451	2452	2453	2454	2455
4630	2456	2457	2458	2459	2460	2461	2462	2463
4640	2464	2465	2466	2467	2468	2469	2470	2471
4650	2472	2473	2474	2475	2476	2477	2478	2479
4660	2480	2481	2482	2483	2484	2485	2486	2487
4670	2488	2489	2490	2491	2492	2493	2494	2495
4700	2496	2497	2498	2499	2500	2501	2502	2503
4710	2504	2505	2506	2507	2508	2509	2510	2511
4720	2512	2513	2514	2515	2516	2517	2518	2519
4730	2520	2521	2522	2523	2524	2525	2526	2527
4740	2528	2529	2530	2531	2532	2533	2534	2535
4750	2536	2537	2538	2539	2540	2541	2542	2543
4760	2544	2545	2546	2547	2548	2549	2550	2551
4770	2552	2553	2554	2555	2556	2557	2558	2559

OCTAL	0	1	2	3	4	5	6	7
5000	2560	2561	2562	2563	2564	2565	2566	2567
5010	2568	2569	2570	2571	2572	2573	2574	2575
5020	2576	2577	2578	2579	2580	2581	2582	2583
5030	2584	2585	2586	2587	2588	2589	2590	2591
5040	2592	2593	2594	2595	2596	2597	2598	2599
5050	2600	2601	2602	2603	2604	2605	2606	2607
5060	2608	2609	2610	2611	2612	2613	2614	2615
5070	2616	2617	2618	2619	2620	2621	2622	2623
5100	2624	2625	2626	2627	2628	2629	2630	2631
5110	2632	2633	2634	2635	2636	2637	2638	2639
5120	2640	2641	2642	2643	2644	2645	2646	2647
5130	2648	2649	2650	2651	2652	2653	2654	2655
5140	2656	2657	2658	2659	2660	2661	2662	2663
5150	2664	2665	2666	2667	2668	2669	2670	2671
5160	2672	2673	2674	2675	2676	2677	2678	2679
5170	2680	2681	2682	2683	2684	2685	2686	2687
5200	2688	2689	2690	2691	2692	2693	2694	2695
5210	2696	2697	2698	2699	2700	2701	2702	2703
5220	2704	2705	2706	2707	2708	2709	2710	2711
5230	2712	2713	2714	2715	2716	2717	2718	2719
5240	2720	2721	2722	2723	2724	2725	2726	2727
5250	2728	2729	2730	2731	2732	2733	2734	2735
5260	2736	2737	2738	2739	2740	2741	2742	2743
5270	2744	2745	2746	2747	2748	2749	2750	2751
5300	2752	2753	2754	2755	2756	2757	2758	2759
5310	2760	2761	2762	2763	2764	2765	2766	2767
5320	2768	2769	2770	2771	2772	2773	2774	2775
5330	2776	2777	2778	2779	2780	2781	2782	2783
5340	2784	2785	2786	2787	2788	2789	2790	2791
5350	2792	2793	2794	2795	2796	2797	2798	2799
5360	2800	2801	2802	2803	2804	2805	2806	2807
5370	2808	2809	2810	2811	2812	2813	2814	2815

OCTAL	0	1	2	3	4	5	6	7
5400	2816	2817	2818	2819	2820	2821	2822	2823
5410	2824	2825	2826	2827	2828	2829	2830	2831
5420	2832	2833	2834	2835	2836	2837	2838	2839
5430	2840	2841	2842	2843	2844	2845	2846	2847
5440	2848	2849	2850	2851	2852	2853	2854	2855
5450	2856	2857	2858	2859	2860	2861	2862	2863
5460	2864	2865	2866	2867	2868	2869	2870	2871
5470	2872	2873	2874	2875	2876	2877	2878	2879
5500	2880	2881	2882	2883	2884	2885	2886	2887
5510	2888	2889	2890	2891	2892	2893	2894	2895
5520	2896	2897	2898	2899	2900	2901	2902	2903
5530	2904	2905	2906	2907	2908	2909	2910	2911
5540	2912	2913	2914	2915	2916	2917	2918	2919
5550	2920	2921	2922	2923	2924	2925	2926	2927
5560	2928	2929	2930	2931	2932	2933	2934	2935
5570	2936	2937	2938	2939	2940	2941	2942	2943
5600	2944	2945	2946	2947	2948	2949	2950	2951
5610	2952	2953	2954	2955	2956	2957	2958	2959
5620	2960	2961	2962	2963	2964	2965	2966	2967
5630	2968	2969	2970	2971	2972	2973	2974	2975
5640	2976	2977	2978	2979	2980	2981	2982	2983
5650	2984	2985	2986	2987	2988	2989	2990	2991
5660	2992	2993	2994	2995	2996	2997	2998	2999
5670	3000	3001	3002	3003	3004	3005	3006	3007
5700	3008	3009	3010	3011	3012	3013	3014	3015
5710	3016	3017	3018	3019	3020	3021	3022	3023
5720	3024	3025	3026	3027	3028	3029	3030	3031
5730	3032	3033	3034	3035	3036	3037	3038	3039
5740	3040	3041	3042	3043	3044	3045	3046	3047
5750	3048	3049	3050	3051	3052	3053	3054	3055
5760	3056	3057	3058	3059	3060	3061	3062	3063
5770	3064	3065	3066	3067	3068	3069	3070	3071

OCTAL	0	1	2	3	4	5	6	7
6000	3072	3073	3074	3075	3076	3077	3078	3079
6010	3080	3081	3082	3083	3084	3085	3086	3087
6020	3088	3089	3090	3091	3092	3093	3094	3095
6030	3096	3097	3098	3099	3100	3101	3102	3103
6040	3104	3105	3106	3107	3108	3109	3110	3111
6050	3112	3113	3114	3115	3116	3117	3118	3119
6060	3120	3121	3122	3123	3124	3125	3126	3127
6070	3128	3129	3130	3131	3132	3133	3134	3135
6100	3136	3137	3138	3139	3140	3141	3142	3143
6110	3144	3145	3146	3147	3148	3149	3150	3151
6120	3152	3153	3154	3155	3156	3157	3158	3159
6130	3160	3161	3162	3163	3164	3165	3166	3167
6140	3168	3169	3170	3171	3172	3173	3174	3175
6150	3176	3177	3178	3179	3180	3181	3182	3183
6160	3184	3185	3186	3187	3188	3189	3190	3191
6170	3192	3193	3194	3195	3196	3197	3198	3199
6200	3200	3201	3202	3203	3204	3205	3206	3207
6210	3208	3209	3210	3211	3212	3213	3214	3215
6220	3216	3217	3218	3219	3220	3221	3222	3223
6230	3224	3225	3226	3227	3228	3229	3230	3231
6240	3232	3233	3234	3235	3236	3237	3238	3239
6250	3240	3241	3242	3243	3244	3245	3246	3247
6260	3248	3249	3250	3251	3252	3253	3254	3255
6270	3256	3257	3258	3259	3260	3261	3262	3263
6300	3264	3265	3266	3267	3268	3269	3270	3271
6310	3272	3273	3274	3275	3276	3277	3278	3279
6320	3280	3281	3282	3283	3284	3285	3286	3287
6330	3288	3289	3290	3291	3292	3293	3294	3295
6340	3296	3297	3298	3299	3300	3301	3302	3303
6350	3304	3305	3306	3307	3308	3309	3310	3311
6360	3312	3313	3314	3315	3316	3317	3318	3319
6370	3320	3321	3322	3323	3324	3325	3326	3327

OCTAL	0	1	2	3	4	5	6	7
6400	3328	3329	3330	3331	3332	3333	3334	3335
6410	3336	3337	3338	3339	3340	3341	3342	3343
6420	3344	3345	3346	3347	3348	3349	3350	3351
6430	3352	3353	3354	3355	3356	3357	3358	3359
6440	3360	3361	3362	3363	3364	3365	3366	3367
6450	3368	3369	3370	3371	3372	3373	3374	3375
6460	3376	3377	3378	3379	3380	3381	3382	3383
6470	3384	3385	3386	3387	3388	3389	3390	3391
6500	3392	3393	3394	3395	3396	3397	3398	3399
6510	3400	3401	3402	3403	3404	3405	3406	3407
6520	3408	3409	3410	3411	3412	3413	3414	3415
6530	3416	3417	3418	3419	3420	3421	3422	3423
6540	3424	3425	3426	3427	3428	3429	3430	3431
6550	3432	3433	3434	3435	3436	3437	3438	3439
6560	3440	3441	3442	3443	3444	3445	3446	3447
6570	3448	3449	3450	3451	3452	3453	3454	3455
6600	3456	3457	3458	3459	3460	3461	3462	3463
6610	3464	3465	3466	3467	3468	3469	3470	3471
6620	3472	3473	3474	3475	3476	3477	3478	3479
6630	3480	3481	3482	3483	3484	3485	3486	3487
6640	3488	3489	3490	3491	3492	3493	3494	3495
6650	3496	3497	3498	3499	3500	3501	3502	3503
6660	3504	3505	3506	3507	3508	3509	3510	3511
6670	3512	3513	3514	3515	3516	3517	3518	3519
6700	3520	3521	3522	3523	3524	3525	3526	3527
6710	3528	3529	3530	3531	3532	3533	3534	3535
6720	3536	3537	3538	3539	3540	3541	3542	3543
6730	3544	3545	3546	3547	3548	3549	3550	3551
6740	3552	3553	3554	3555	3556	3557	3558	3559
6750	3560	3561	3562	3563	3564	3565	3566	3567
6760	3568	3569	3570	3571	3572	3573	3574	3575
6770	3576	3577	3578	3579	3580	3581	3582	3583

OCTAL	0	1	2	3	4	5	6	7
7000	3584	3585	3586	3587	3588	3589	3590	3591
7010	3592	3593	3594	3595	3596	3597	3598	3599
7020	3600	3601	3602	3603	3604	3605	3606	3607
7030	3608	3609	3610	3611	3612	3613	3614	3615
7040	3616	3617	3618	3619	3620	3621	3622	3623
7050	3624	3625	3626	3627	3628	3629	3630	3631
7060	3632	3633	3634	3635	3636	3637	3638	3639
7070	3640	3641	3642	3643	3644	3645	3646	3647
7100	3648	3649	3650	3651	3652	3653	3654	3655
7110	3656	3657	3658	3659	3660	3661	3662	3663
7120	3664	3665	3666	3667	3668	3669	3670	3671
7130	3672	3673	3674	3675	3676	3677	3678	3679
7140	3680	3681	3682	3683	3684	3685	3686	3687
7150	3688	3689	3690	3691	3692	3693	3694	3695
7160	3696	3697	3698	3699	3700	3701	3702	3703
7170	3704	3705	3706	3707	3708	3709	3710	3711
7200	3712	3713	3714	3715	3716	3717	3718	3719
7210	3720	3721	3722	3723	3724	3725	3726	3727
7220	3728	3729	3730	3731	3732	3733	3734	3735
7230	3736	3737	3738	3739	3740	3741	3742	3743
7240	3744	3745	3746	3747	3748	3749	3750	3751
7250	3752	3753	3754	3755	3756	3757	3758	3759
7260	3760	3761	3762	3763	3764	3765	3766	3767
7270	3768	3769	3770	3771	3772	3773	3774	3775
7300	3776	3777	3778	3779	3780	3781	3782	3783
7310	3784	3785	3786	3787	3788	3789	3790	3791
7320	3792	3793	3794	3795	3796	3797	3798	3799
7330	3800	3801	3802	3803	3804	3805	3806	3807
7340	3808	3809	3810	3811	3812	3813	3814	3815
7350	3816	3817	3818	3819	3820	3821	3822	3823
7360	3824	3825	3826	3827	3828	3829	3830	3831
7370	3832	3833	3834	3835	3836	3837	3838	3839

OCTAL	0	1	2	3	4	5	6	7
7400	3840	3841	3842	3843	3844	3845	3846	3847
7410	3848	3849	3850	3851	3852	3853	3854	3855
7420	3856	3857	3858	3859	3860	3861	3862	3863
7430	3864	3865	3866	3867	3868	3869	3870	3871
7440	3872	3873	3874	3875	3876	3877	3878	3879
7450	3880	3881	3882	3883	3884	3885	3886	3887
7460	3888	3889	3890	3891	3892	3893	3894	3895
7470	3896	3897	3898	3899	3900	3901	3902	3903
7500	3904	3905	3906	3907	3908	3909	3910	3911
7510	3912	3913	3914	3915	3916	3917	3918	3919
7520	3920	3921	3922	3923	3924	3925	3926	3927
7530	3928	3929	3930	3931	3932	3933	3934	3935
7540	3936	3937	3938	3939	3940	3941	3942	3943
7550	3944	3945	3946	3947	3948	3949	3950	3951
7560	3952	3953	3954	3955	3956	3957	3958	3959
7570	3960	3961	3962	3963	3964	3965	3966	3967
7600	3968	3969	3970	3971	3972	3973	3974	3975
7610	3976	3977	3978	3979	3980	3981	3982	3983
7620	3984	3985	3986	3987	3988	3989	3990	3991
7630	3992	3993	3994	3995	3996	3997	3998	3999
7640	4000	4001	4002	4003	4004	4005	4006	4007
7650	4008	4009	4010	4011	4012	4013	4014	4015
7660	4016	4017	4018	4019	4020	4021	4022	4023
7670	4024	4025	4026	4027	4028	4029	4030	4031
7700	4032	4033	4034	4035	4036	4037	4038	4039
7710	4040	4041	4042	4043	4044	4045	4046	4047
7720	4048	4049	4050	4051	4052	4053	4054	4055
7730	4056	4057	4058	4059	4060	4061	4062	4063
7740	4064	4065	4066	4067	4068	4069	4070	4071
7750	4072	4073	4074	4075	4076	4077	4078	4079
7760	4080	4081	4082	4083	4084	4085	4086	4087
7770	4088	4089	4090	4091	4092	4093	4094	4095

Octal	Decimal
10000	4096
20000	8192
30000	12288
40000	16384
50000	20480
60000	24576
70000	28672

C Squares and Square Roots of Numbers From 1 To 1000

N	N^2	$\sqrt{N}$	N	N^2	$\sqrt{N}$
1	1	1.000	41	16 81	6.403
2	4	1.414	42	17 64	6.481
3	9	1.732	43	18 49	6.557
4	16	2.000	44	19 36	6.633
5	25	2.236	45	20 25	6.708
6	36	2.449	46	21 16	6.782
7	49	2.646	47	22 09	6.856
8	64	2.828	48	23 04	6.928
9	81	3.000	49	24 01	7.000
10	100	3.162	50	25 00	7.071
11	1 21	3.317	51	26 01	7.141
12	1 44	3.464	52	27 04	7.211
13	1 69	3.606	53	28 09	7.280
14	1 96	3.742	54	29 16	7.348
15	2 25	3.873	55	30 25	7.416
16	2 56	4.000	56	31 36	7.483
17	2 89	4.123	57	32 49	7.550
18	3 24	4.243	58	33 64	7.616
19	3 61	4.359	59	34 81	7.681
20	4 00	4.472	60	36 00	7.746
21	4 41	4.583	61	37 21	7.810
22	4 84	4.690	62	38 44	7.874
23	5 29	4.796	63	39 69	7.937
24	5 76	4.899	64	40 96	8.000
25	6 25	5.000	65	42 25	8.062
26	6 76	5.099	66	43 56	8.124
27	7 29	5.196	67	44 89	8.185
28	7 84	5.292	68	46 24	8.246
29	8 41	5.385	69	47 61	8.307
30	9 00	5.477	70	49 00	8.367
31	9 61	5.568	71	50 41	8.426
32	10 24	5.657	72	51 84	8.485
33	10 89	5.745	73	53 29	8.544
34	11 56	5.831	74	54 76	8.602
35	12 25	5.916	75	56 25	8.660
36	12 96	6.000	76	57 76	8.718
37	13 69	6.083	77	59 29	8.775
38	14 44	6.164	78	60 84	8.832
39	15 21	6.245	79	62 41	8.888
40	16 00	6.325	80	64 00	8.944

N	N^2	$\sqrt{N}$	N	N^2	$\sqrt{N}$
81	65 61	9.000	121	1 46 41	11.000
82	67 24	9.055	122	1 48 84	11.045
83	68 89	9.110	123	1 51 29	11.091
84	70 56	9.165	124	1 53 76	11.136
85	72 25	9.220	125	1 56 25	11.180
86	73 96	9.274	126	1 58 76	11.225
87	75 69	9.327	127	1 61 29	11.269
88	77 44	9.381	128	1 63 84	11.314
89	79 21	9.434	129	1 66 41	11.358
90	81 00	9.487	130	1 69 00	11.402
91	82 81	9.539	131	1 71 61	11.446
92	84 64	9.592	132	1 74 24	11.489
93	86 49	9.644	133	1 76 89	11.533
94	88 36	9.695	134	1 79 56	11.576
95	90 25	9.747	135	1 82 25	11.619
96	92 16	9.798	136	1 84 96	11.662
97	94 09	9.849	137	1 87 69	11.705
98	96 04	9.899	138	1 90 44	11.747
99	98 01	9.950	139	1 93 21	11.790
100	1 00 00	10.000	140	1 96 00	11.832
101	1 02 01	10.050	141	1 98 81	11.874
102	1 04 04	10.100	142	2 01 64	11.916
103	1 06 09	10.149	143	2 04 49	11.958
104	1 08 16	10.198	144	2 07 36	12.000
105	1 10 25	10.247	145	2 10 25	12.042
106	1 12 36	10.296	146	2 13 16	12.083
107	1 14 49	10.344	147	2 16 09	12.124
108	1 16 64	10.392	148	2 19 04	12.166
109	1 18 81	10.440	149	2 22 01	12.207
110	1 21 00	10.488	150	2 25 00	12.247
111	1 23 21	10.536	151	2 28 01	12.288
112	1 25 44	10.583	152	2 31 04	12.329
113	1 27 69	10.630	153	2 34 09	12.369
114	1 29 96	10.677	154	2 37 16	12.410
115	1 32 25	10.724	155	2 40 25	12.450
116	1 34 56	10.770	156	2 43 36	12.490
117	1 36 89	10.817	157	2 46 49	12.530
118	1 39 24	10.863	158	2 49 64	12.570
119	1 41 61	10.909	159	2 52 81	12.610
120	1 44 00	10.954	160	2 56 00	12.649

N	N^2	$\sqrt{N}$	N	N^2	$\sqrt{N}$
161	2 59 21	12.689	201	4 04 01	14.177
162	2 62 44	12.728	202	4 08 04	14.213
163	2 65 69	12.767	203	4 12 09	14.248
164	2 68 96	12.806	204	4 16 16	14.283
165	2 72 25	12.845	205	4 20 25	14.318
166	2 75 56	12.884	206	4 24 36	14.353
167	2 78 89	12.923	207	4 28 49	14.387
168	2 82 24	12.961	208	4 32 64	14.422
169	2 85 61	13.000	209	4 36 81	14.457
170	2 89 00	13.038	210	4 41 00	14.491
171	2 92 41	13.077	211	4 45 21	14.526
172	2 95 84	13.115	212	4 49 44	14.560
173	2 99 29	13.153	213	4 53 69	14.595
174	3 02 76	13.191	214	4 57 96	14.629
175	3 06 25	13.229	215	4 62 25	14.663
176	3 09 76	13.266	216	4 66 56	14.697
177	3 13 29	13.304	217	4 70 89	14.731
178	3 16 84	13.342	218	4 75 24	14.765
179	3 20 41	13.379	219	4 79 61	14.799
180	3 24 00	13.416	220	4 84 00	14.832
181	3 27 61	13.454	221	4 88 41	14.866
182	3 31 24	13.491	222	4 92 84	14.900
183	3 34 89	13.528	223	4 97 29	14.933
184	3 38 56	13.565	224	5 01 76	14.967
185	3 42 25	13.601	225	5 06 25	15.000
186	3 45 96	13.638	226	5 10 76	15.033
187	3 49 69	13.675	227	5 15 29	15.067
188	3 53 44	13.711	228	5 19 84	15.100
189	3 57 21	13.748	229	5 24 41	15.133
190	3 61 00	13.784	230	5 29 00	15.166
191	3 64 81	13.820	231	5 33 61	15.199
192	3 68 64	13.856	232	5 38 24	15.232
193	3 72 49	13.892	233	5 42 89	15.264
194	3 76 36	13.928	234	5 47 56	15.297
195	3 80 25	13.964	235	5 52 25	15.330
196	3 84 16	14.000	236	5 56 96	15.362
197	3 88 09	14.036	237	5 61 69	15.395
198	3 92 04	14.071	238	5 66 44	15.427
199	3 96 01	14.107	239	5 71 21	15.460
200	4 00 00	14.142	240	5 76 00	15.492

N	N^2	$\sqrt{N}$	N	N^2	$\sqrt{N}$
241	5 80 81	15.524	281	7 89 61	16.763
242	5 85 64	15.556	282	7 95 24	16.793
243	5 90 49	15.588	283	8 00 89	16.823
244	5 95 36	15.620	284	8 06 56	16.852
245	6 00 25	15.652	285	8 12 25	16.882
246	6 05 16	15.684	286	8 17 96	16.912
247	6 10 09	15.716	287	8 23 69	16.941
248	6 15 04	15.748	288	8 29 44	16.971
249	6 20 01	15.780	289	8 35 21	17.000
250	6 25 00	15.811	290	8 41 00	17.029
251	6 30 01	15.843	291	8 46 81	17.059
252	6 35 04	15.875	292	8 52 64	17.088
253	6 40 09	15.906	293	8 58 49	17.117
254	6 45 16	15.937	294	8 64 36	17.146
255	6 50 25	15.969	295	8 70 25	17.176
256	6 55 36	16.000	296	8 76 16	17.205
257	6 60 49	16.031	297	8 82 09	17.234
258	6 65 64	16.062	298	8 88 04	17.263
259	6 70 81	16.093	299	8 94 01	17.292
260	6 76 00	16.125	300	9 00 00	17.321
261	6 81 21	16.155	301	9 06 01	17.349
262	6 86 44	16.186	302	9 12 04	17.378
263	6 91 69	16.217	303	9 18 09	17.407
264	6 96 96	16.248	304	9 24 16	17.436
265	7 02 25	16.279	305	9 30 25	17.464
266	7 07 56	16.310	306	9 36 36	17.493
267	7 12 89	16.340	307	9 42 49	17.521
268	7 18 24	16.371	308	9 48 64	17.550
269	7 23 61	16.401	309	9 54 81	17.578
270	7 29 00	16.432	310	9 61 00	17.607
271	7 34 41	16.462	311	9 67 21	17.635
272	7 39 84	16.492	312	9 73 44	17.664
273	7 45 29	16.523	313	9 79 69	17.692
274	7 50 76	16.553	314	9 85 96	17.720
275	7 56 25	16.583	315	9 92 25	17.748
276	7 61 76	16.613	316	9 98 56	17.776
277	7 67 29	16.643	317	10 04 89	17.804
278	7 72 84	16.673	318	10 11 24	17.833
279	7 78 41	16.703	319	10 17 61	17.861
280	7 84 00	16.733	320	10 24 00	17.889

N	N^2	$\sqrt{N}$	N	N^2	$\sqrt{N}$
321	10 30 41	17.916	361	13 03 21	19.000
322	10 36 84	17.944	362	13 10 44	19.026
323	10 43 29	17.972	363	13 17 69	19.053
324	10 49 76	18.000	364	13 24 96	19.079
325	10 56 25	18.028	365	13 32 25	19.105
326	10 62 76	18.055	366	13 39 56	19.131
327	10 69 29	18.083	367	13 46 89	19.157
328	10 75 84	18.111	368	13 54 24	19.183
329	10 82 41	18.138	369	13 61 61	19.209
330	10 89 00	18.166	370	13 69 00	19.235
331	10 95 61	18.193	371	13 76 41	19.261
332	11 02 24	18.221	372	13 83 84	19.287
333	11 08 89	18.248	373	13 91 29	19.313
334	11 15 56	18.276	374	13 98 76	19.339
335	11 22 25	18.303	375	14 06 25	19.365
336	11 28 96	18.330	376	14 13 76	19.391
337	11 35 69	18.358	377	14 21 29	19.416
338	11 42 44	18.385	378	14 28 84	19.442
339	11 49 21	18.412	379	14 36 41	19.468
340	11 56 00	18.439	380	14 44 00	19.494
341	11 62 81	18.466	381	14 51 61	19.519
342	11 69 64	18.493	382	14 59 24	19.545
343	11 76 49	18.520	383	14 66 89	19.570
344	11 83 36	18.547	384	14 74 56	19.596
345	11 90 25	18.574	385	14 82 25	19.621
346	11 97 16	18.601	386	14 89 96	19.647
347	12 04 09	18.628	387	14 97 69	19.672
348	12 11 04	18.655	388	15 05 44	19.698
349	12 18 01	18.682	389	15 13 21	19.723
350	12 25 00	18.708	390	15 21 00	19.748
351	12 32 01	18.735	391	15 28 81	19.774
352	12 39 04	18.762	392	15 36 64	19.799
353	12 46 09	18.788	393	15 44 49	19.824
354	12 53 16	18.815	394	15 52 36	19.849
355	12 60 25	18.841	395	15 60 25	19.875
356	12 67 36	18.868	396	15 68 16	19.900
357	12 74 49	18.894	397	15 76 09	19.925
358	12 81 64	18.921	398	15 84 04	19.950
359	12 88 81	18.947	399	15 92 01	19.975
360	12 96 00	18.974	400	16 00 00	20.000

N	N^2	$\sqrt{N}$	N	N^2	$\sqrt{N}$
401	16 08 01	20.025	441	19 44 81	21.000
402	16 16 04	20.050	442	19 53 64	21.024
403	16 24 09	20.075	443	19 62 49	21.048
404	16 32 16	20.100	444	19 71 36	21.071
405	16 40 25	20.125	445	19 80 25	21.095
406	16 48 36	20.149	446	19 89 16	21.119
407	16 56 49	20.174	447	19 98 09	21.142
408	16 64 64	20.199	448	20 07 04	21.166
409	16 72 81	20.224	449	20 16 01	21.190
410	16 81 00	20.248	450	20 25 00	21.213
411	16 89 21	20.273	451	20 34 01	21.237
412	16 97 44	20.298	452	20 43 04	21.260
413	17 05 69	20.322	453	20 52 09	21.284
414	17 13 96	20.347	454	20 61 16	21.307
415	17 22 25	20.372	455	20 70 25	21.331
416	17 30 56	20.396	456	20 79 36	21.354
417	17 38 89	20.421	457	20 88 49	21.378
418	17 47 24	20.445	458	20 97 64	21.401
419	17 55 61	20.469	459	21 06 81	21.424
420	17 64 00	20.494	460	21 16 00	21.448
421	17 72 41	20.518	461	21 25 21	21.471
422	17 80 84	20.543	462	21 34 44	21.494
423	17 89 29	20.567	463	2I 43 69	21.517
424	17 97 76	20.591	464	21 52 96	21.541
425	18 06 25	20.616	465	21 62 25	21.564
426	18 14 76	20.640	466	2 I 71 56	21.587
427	18 23 29	20.664	467	21 80 89	21.610
428	18 31 84	20.688	468	21 90 24	21.633
429	18 40 41	20.712	469	21 99 61	21.656
430	18 49 00	20.736	470	22 09 00	21.679
431	18 57 61	20.761	471	22 18 41	21.703
432	18 66 24	20.785	472	22 27 84	21.726
433	18 74 89	20.809	473	22 37 29	21.749
434	18 83 56	20.833	474	22 46 76	21.772
435	18 92 25	20.857	475	22 56 25	21.794
436	19 00 96	20.881	476	22 65 76	21.817
437	19 09 69	20.905	477	22 75 29	21.840
438	19 18 44	20.928	478	22 84 84	21.863
439	19 27 21	20.952	479	22 94 41	21.886
440	19 36 00	20.976	480	23 04 00	21.909

N	N^2	$\sqrt{N}$	N	N^2	$\sqrt{N}$
481	23 13 61	21.932	521	27 14 41	22.825
482	23 23 24	21.954	522	27 24 84	22.847
483	23 32 89	21.977	523	27 35 29	22.869
484	23 42 56	22.000	524	27 45 76	22.891
485	23 52 25	22.023	525	27 56 25	22.913
486	23 61 96	22.045	526	27 66 76	22.935
487	23 71 69	22.068	527	27 77 29	22.956
488	23 81 44	22.091	528	27 87 84	22.978
489	23 91 21	22.113	529	27 98 41	23.000
490	24 01 00	22.136	530	28 09 00	23.022
491	24 10 81	22.159	531	28 19 61	23.043
492	24 20 64	22.181	532	28 30 24	23.065
493	24 30 49	22.204	533	28 40 89	23.087
494	24 40 36	22.226	534	28 51 56	23.108
495	24 50 25	22.249	535	28 62 25	23.130
496	24 60 16	22.271	536	28 72 96	23.152
497	24 70 09	22.293	537	28 83 69	23.173
498	24 80 04	22.316	538	28 94 44	23.195
499	24 90 01	22.338	539	29 05 21	23.216
500	25 00 00	22.361	540	29 16 00	23.238
501	25 10 01	22.383	541	29 26 81	23.259
502	25 20 04	22.405	542	29 37 64	23.281
503	25 30 09	22.428	543	29 48 49	23.302
504	25 40 16	22.450	544	29 59 36	23.324
505	25 50 25	22.472	545	29 70 25	23.345
506	25 60 36	22.494	546	29 81 16	23.367
507	25 70 49	22.517	547	29 92 09	23.388
508	25 80 64	22.539	548	30 03 04	23.409
509	25 90 81	22.561	549	30 14 01	23.431
510	26 01 00	22.583	550	30 25 00	23.452
511	26 11 21	22.605	551	30 36 01	23.473
512	26 21 44	22.627	552	30 47 04	23.495
513	26 31 69	22.650	553	30 58 09	23.516
514	26 41 96	22.672	554	30 69 16	23.537
515	26 52 25	22.694	555	30 80 25	23.558
516	26 62 56	22.716	556	30 91 36	23.580
517	26 72 89	22.738	557	31 02 49	23.601
518	26 83 24	22.760	558	31 13 64	23.622
519	26 93 61	22.782	559	31 24 81	23.643
520	27 04 00	22.804	560	31 36 00	23.664

N	N^2	$\sqrt{N}$	N	N^2	$\sqrt{N}$
561	31 47 21	23.685	601	36 12 01	24.515
562	31 58 44	23.707	602	36 24 04	24.536
563	31 69 69	23.728	603	36 36 09	24.556
564	31 80 96	23.749	604	36 48 16	24.576
565	31 92 25	23.770	605	36 60 25	24.597
566	32 03 56	23.791	606	36 72 36	24.617
567	32 14 89	23.812	607	36 84 49	24.637
568	32 26 24	23.833	608	36 96 64	24.658
569	32 37 61	23.854	609	37 08 81	24.678
570	32 49 00	23.875	610	37 21 00	24.698
571	32 60 41	23.896	611	37 33 21	24.718
572	32 71 84	23.917	612	37 45 44	24.739
573	32 83 29	23.937	613	37 57 69	24.759
574	32 94 76	23.958	614	37 69 96	24.779
575	33 06 25	23.979	615	37 82 25	24.799
576	33 17 76	24.000	616	37 94 56	24.819
577	33 29 29	24.021	617	38 06 89	24.839
578	33 40 84	24.042	618	38 19 24	24.860
579	33 52 41	24.062	619	38 31 61	24.880
580	33 64 00	24.083	620	38 44 00	24.900
581	33 75 61	24.104	621	38 56 41	24.920
582	33 87 24	24.125	622	38 68 84	24.940
583	33 98 89	24.145	623	38 81 29	24.960
584	34 10 56	24.166	624	38 93 76	24.980
585	34 22 25	24.187	625	39 06 25	25.000
586	34 33 96	24.207	626	39 18 76	25.020
587	34 45 69	24.228	627	39 31 29	25.040
588	34 57 44	24.249	628	39 43 84	25.060
589	34 69 21	24.269	629	39 56 41	25.080
590	34 81 00	24.290	630	39 69 00	25.100
591	34 92 81	24.310	631	39 81 61	25.120
592	35 04 64	24.331	632	39 94 24	25.140
593	35 16 49	24.352	633	40 06 89	25.159
594	35 28 36	24.372	634	40 19 56	25.179
595	35 40 25	24.393	635	40 32 25	25.199
596	35 52 16	24.413	636	40 44 96	25.219
597	35 64 09	24.434	637	40 57 69	25.239
598	35 76 04	24.454	638	40 70 44	25.259
599	35 88 01	24.474	639	40 83 21	25.278
600	36 00 00	24.495	640	40 96 00	25.298

N	N^2	$\sqrt{N}$	N	N^2	$\sqrt{N}$
641	41 08 81	25.318	681	46 37 61	26.096
642	41 21 64	25.338	682	46 51 24	26.115
643	41 34 49	25.357	683	46 64 89	26.134
644	41 47 36	25.377	684	46 78 56	26.153
645	41 60 25	25.397	685	46 92 25	26.173
646	41 73 16	25.417	686	47 05 96	26.192
647	41 86 09	25.436	687	47 19 69	26.211
648	41 99 04	25.456	688	47 33 44	26.230
649	42 12 01	25.475	689	47 47 21	26.249
650	42 25 00	25.495	690	47 61 00	26.268
651	42 38 01	25.515	691	47 74 81	26.287
652	42 51 04	25.534	692	47 88 64	26.306
653	42 64 09	25.554	693	48 02 49	26.325
654	42 77 16	25.573	694	48 16 36	26.344
655	42 90 25	25.593	695	48 30 25	26.363
656	43 03 36	25.612	696	48 44 16	26.382
657	43 16 49	25.632	697	48 58 09	26.401
658	43 29 64	25.652	698	48 72 04	26.420
659	43 42 81	25.671	699	48 86 01	26.439
660	43 56 00	25.690	700	49 00 00	26.458
661	43 69 21	25.710	701	49 14 01	26.476
662	42 82 44	25.729	702	49 28 04	26.495
663	43 95 69	25.749	703	49 42 09	26.514
664	44 08 96	25.768	704	49 56 16	26.533
665	44 22 25	25.788	705	49 70 25	26.552
666	44 35 56	25.807	706	49 84 36	26.571
667	44 48 89	25.826	707	49 98 49	26.589
668	44 62 24	25.846	708	50 12 64	26.608
669	44 75 61	25.865	709	50 26 81	26.627
670	44 89 00	25.884	710	50 41 00	26.646
671	45 02 41	25.904	711	50 55 21	26.665
672	45 15 84	25.923	712	50 69 44	26.683
673	45 29 29	25.942	713	50 83 69	26.702
674	45 42 76	25.962	714	50 97 96	26.721
675	45 56 25	25.981	715	51 12 25	26.739
676	45 69 76	26.000	716	51 26 56	26.758
677	45 83 29	26.019	717	51 40 89	26.777
678	45 96 84	26.038	718	51 55 24	26.796
679	46 10 41	26.058	719	51 69 61	26.814
680	46 24 00	26.077	720	51 84 00	26.833

N	N^2	$\sqrt{N}$	N	N^2	$\sqrt{N}$
721	51 98 41	26.851	761	57 91 21	27.586
722	52 12 84	26.870	762	58 06 44	27.604
723	52 27 29	26.889	763	58 21 69	27.622
724	52 41 76	26.907	764	58 36 96	27.641
725	52 56 25	26.926	765	58 52 25	27.659
726	52 70 76	26.944	766	58 67 56	27.677
727	52 85 29	26.963	767	58 82 89	27.695
728	52 99 84	26.981	768	58 98 24	27.713
729	53 14 41	27.000	769	59 13 61	27.731
730	53 29 00	27.019	770	59 29 00	27.749
731	53 43 61	27.037	771	59 44 41	27.767
732	53 58 24	27.055	772	59 59 84	27.785
733	53 72 89	27.074	773	59 75 29	27.803
734	53 87 56	27.092	774	59 90 76	27.821
735	54 02 25	27.111	775	60 06 25	27.839
736	54 16 96	27.129	776	60 21 76	27.857
737	54 31 69	27.148	777	60 37 29	27.875
738	54 46 44	27.166	778	60 52 84	27.893
739	54 61 21	27.185	779	60 68 41	27.911
740	54 76 00	27.203	780	60 84 00	27.928
741	54 90 81	27.221	781	60 99 61	27.946
742	55 05 64	27.240	782	61 15 24	27.964
743	55 20 49	27.258	783	61 30 89	27.982
744	55 35 36	27.276	784	61 46 56	28.000
745	55 50 25	27.295	785	61 62 25	28.018
746	55 65 16	27.313	786	61 77 96	28.036
747	55 80 09	27.331	787	61 93 69	28.054
748	55 95 04	27.350	788	62 09 44	28.071
749	56 10 01	27.368	789	62 25 21	28.089
750	56 25 00	27.386	790	62 41 00	28.107
751	56 40 01	27.404	791	62 56 81	28.125
752	56 55 04	27.423	792	62 72 64	28.142
753	56 70 09	27.441	793	62 88 49	28.160
754	56 85 16	27.459	794	63 04 36	28.178
755	57 00 25	27.477	795	63 20 25	28.196
756	57 15 36	27.495	796	63 36 16	28.213
757	57 30 49	27.514	797	63 52 09	28.231
758	57 45 64	27.532	798	63 68 04	28.249
759	57 60 81	27.550	799	63 84 01	28.267
760	57 76 00	27.568	800	64 00 00	28.284

N	N^2	$\sqrt{N}$	N	N^2	$\sqrt{N}$
801	64 16 01	28.302	841	70 72 81	29.000
802	64 32 04	28.320	842	70 89 64	29.017
803	64 48 09	28.337	843	71 06 49	29.034
804	64 64 16	28.355	844	71 23 36	29.052
805	64 80 25	28.373	845	71 40 25	29.069
806	64 96 36	28.390	846	71 57 16	29.086
807	65 12 49	28.408	847	71 74 09	29.103
808	65 28 64	28.425	848	71 91 04	29.120
809	65 44 81	28.443	849	72 08 01	29.138
810	65 61 00	28.460	850	72 25 00	29.155
811	65 77 21	28.478	851	72 42 01	29.172
812	65 93 44	28.496	852	72 59 04	29.189
813	66 09 69	28.513	853	72 76 09	29.206
814	66 25 96	28.531	854	72 93 16	29.223
815	66 42 25	28.548	855	73 10 25	29.240
816	66 58 56	28.566	856	73 27 36	29.257
817	66 74 89	28.583	857	73 44 49	29.275
818	66 91 24	28.601	858	73 61 64	29.292
819	67 07 61	28.618	859	73 78 81	29.309
820	67 24 00	28.636	860	73 96 00	29.326
821	67 40 41	28.653	861	74 13 21	29.343
822	67 56 84	23.671	862	74 30 44	29.360
823	67 73 29	28.688	863	74 47 69	29.377
824	67 89 76	28.705	864	74 64 96	29.394
825	68 06 25	28.723	865	74 82 25	29.411
826	68 22 76	28.740	866	74 99 56	29.428
827	68 39 29	28.758	867	75 16 89	29.445
828	68 55 84	28.775	868	75 34 24	29.462
829	68 72 41	28.792	869	75 51 61	29.479
830	68 89 00	28.810	870	75 69 00	29.496
831	69 05 61	28.827	871	75 86 41	29.513
832	69 22 24	28.844	872	76 03 84	29.530
833	69 38 89	28.862	873	76 21 29	29.547
834	69 55 56	28.879	874	76 38 76	29.563
835	69 72 25	28.896	875	76 56 25	29.580
836	69 88 96	28.914	876	76 73 76	29.597
837	70 05 69	28.931	877	76 91 29	29.614
838	70 22 44	28.948	878	77 08 84	29.631
839	70 39 21	28.965	879	77 26 41	29.648
840	70 56 00	23.983	880	77 44 00	29.665

N	N^2	$\sqrt{N}$	N	N^2	$\sqrt{N}$
881	77 61 61	29.682	921	84 82 41	30.348
882	77 79 24	29.698	922	85 00 84	30.364
883	77 96 89	29.715	923	85 19 29	30.381
884	78 14 56	29.732	924	85 37 76	30.397
885	78 32 25	29.749	925	85 56 25	30.414
886	78 49 96	29.766	926	85 74 76	30.430
887	78 67 69	29.783	927	85 93 29	30.447
888	78 85 44	29.799	928	86 11 84	30.463
889	79 03 21	29.816	929	86 30 41	30.480
890	79 21 00	29.833	930	86 49 00	30.496
891	79 38 81	29.850	931	86 67 61	30.512
892	79 56 64	29.866	932	86 86 24	30.529
893	79 74 49	29.883	933	87 04 89	30.545
894	79 92 36	29.900	934	87 23 56	30.561
895	80 10 25	29.916	935	87 42 25	30.578
896	80 28 16	29.933	936	87 60 96	30.594
897	80 46 09	29.950	937	87 79 69	30.610
898	80 64 04	29.967	938	87 98 44	30.627
899	80 82 01	29.983	939	88 17 21	30.643
900	81 00 00	30.000	940	88 36 00	30.659
901	81 18 01	30.017	941	88 54 81	30.676
902	81 36 04	30.033	942	88 73 64	30.692
903	81 54 09	30.050	943	88 92 49	30.708
904	81 72 16	30.067	944	89 11 36	30.725
905	81 90 25	30.083	945	89 30 25	30.741
906	82 08 36	30.100	946	89 49 16	30.757
907	82 26 49	30.116	947	89 68 09	30.773
908	82 44 64	30.133	948	89 87 04	30.790
909	82 62 81	30.150	949	90 06 01	30.806
910	82 81 00	30.166	950	90 25 00	30.822
911	82 99 21	30.183	951	90 44 01	30.838
912	83 17 44	30.199	952	90 63 04	30.854
913	83 35 69	30.216	953	90 82 09	30.871
914	83 53 96	30.232	954	91 01 16	30.887
915	83 72 25	30.249	955	91 20 25	30.903
916	83 90 56	30.265	956	91 39 36	30.919
917	84 08 89	30.282	957	91 58 49	30.935
918	84 27 24	30.299	958	91 77 64	30.952
919	84 45 61	30.315	959	91 96 81	30.968
920	84 64 00	30.332	960	92 16 00	30.984

N	N^2	$\sqrt{N}$	N	N^2	$\sqrt{N}$
961	92 35 21	31.000	981	96 23 61	31.321
962	92 54 44	31.016	982	96 43 24	31.337
963	92 73 69	31.032	983	96 62 89	31.353
964	92 92 96	31.048	984	96 82 56	31.369
965	93 12 25	31.064	985	97 02 25	31.385
966	93 31 56	31.081	986	97 21 96	31.401
967	93 50 89	31.097	987	97 41 69	31.417
968	93 70 24	31.113	988	97 61 44	31.432
969	93 89 61	31.129	989	97 81 21	31.448
970	94 09 00	31.145	990	98 01 00	31.464
971	94 28 41	31.161	991	98 20 81	31.480
972	94 47 84	31.177	992	98 40 64	31.496
973	94 67 29	31.193	993	98 60 49	31.512
974	94 86 76	31.209	994	98 80 36	31.528
975	95 06 25	31.225	995	99 00 25	31.544
976	95 25 76	31.241	996	99 20 16	31.559
977	95 45 29	31.257	997	99 40 09	31.575
978	95 64 84	31.273	998	99 60 04	31.591
979	95 84 41	31.289	999	99 80 01	31.607
980	96 04 00	31.305	1000	100 00 00	31.623

D Symbols and Definitions

1. Review of Symbols

Chapter 1

$\{x, y\}$ (braces)	are placed on either side of elements of a set
. . .	indicates an infinite set
$\{\ \ \}$ or $\varnothing$	null or empty set
$\in$	is an element of a set
$\notin$	is not an element of a set
$\subset$	is a subset of a set
$=$	set X contains the following elements; two sets are identical
$\Longleftrightarrow$	indicates equivalence, i.e., two or more sets are equivalent
$\cup$	union of two sets
$\cap$	intersection of two sets
$\overline{A}$ (bar over symbol)	complement of a universal set

Chapter 2

Rules for binary addition:

$$0 + 0 = 0$$
$$0 + 1 = 1$$
$$1 + 0 = 1$$
$$1 + 1 = 0$$

with a carry of 1 to the left

Rules for binary subtraction:

$$0 - 0 = 0$$
$$1 - 1 = 0$$
$$1 - 0 = 1$$
$$0 - 1 = 1$$

with 1 borrowed from the left

Chapter 3

Arithmetic operators:

$+$	add
$-$	subtract
$\times$	multiply
$\div$	divide
A^0, A^1	exponent (superscript)

Logical operators:

$\cdot$	AND⎫ These two are not to be mistaken for
$+$	OR ⎬ the arithmetic operators for multiply and add.
$-$	NOT

Relational operators:

$>$	greater than
$<$	less than
$=$	equal
$\neq$	unequal
$\geq$	greater than or equal to
$\leq$	less than or equal to
Σ	sigma (summation)

Chapter 4

battery (source of power)

light bulb (load)

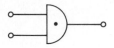

switches (open—closed)

•

+

Boolean AND
Boolean OR

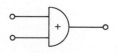

symbolic AND gate

symbolic OR gate

inverter

$\overline{A}$

bar over a symbol signifies the opposite of an existing state (NOT or complement)

NOT Truth Table

$$\overline{0} = 1$$
$$\overline{1} = 0$$

AND Truth Table

$$0 \cdot 0 = 0$$
$$0 \cdot 1 = 0$$
$$1 \cdot 0 = 0$$
$$1 \cdot 1 = 1$$

OR Truth Table

$$0 + 0 = 0$$
$$0 + 1 = 1$$
$$1 + 0 = 1$$
$$1 + 1 = 1$$

Chapter 5

$\supset$	if . . . then . . .
$\equiv$	if and only if (biconditional) symbol for equivalence

decision block

processing block

input/output (I/O) block

terminal block

:	compare
$\neq$	unequal
$\geqslant$	greater than or equal to
$\leqslant$	less than or equal to
$\longrightarrow$	implies $(A \longrightarrow B)$

Chapter 7

D	determinant
$|A|$	determinant of matrix A
A^T	transpose of matrix A

2. Review of Terminology

All terms defined in this appendix are also listed in the index with the applicable page references attached.

Chapter 1

The complement of a set contains all elements of the universal set not included in the original set.

The intersection of two sets is all elements that are common to both sets.

A set may contain some of the elements of another set and other non-related elements.

Sets may be:

Disjoint	have no elements in common
Equivalent	one-to-one relationship between sets
Finite	have definite limits
Identical	two (or more) sets contain exactly the same elements
Infinite	endless
Null	have no elements
Universal	the total environment within which a set (or sets) exists

A subset contains some of the elements of a larger set.

The union of two sets is obtained when every element of two sets is contained in a third set.

Chapter 2

Base (radix)	of a number system is the number of symbols it has, including zero
Binary	base-two number system
Bit	binary digit
Byte	eight bits in consecutive sequence
Hexadecimal	base-sixteen number system
Octal	base-eight number system
Parity bit	an extra bit added to a computer code to aid in verifying accuracy
Quantity	refers to both a unit and a number of units
Subscript	a small number below a symbol or number, indicating the number system (base) being represented (e.g., 45_8, 150_{10})
Superscript	small number above a symbol or number, indicating the power of that quantity (e.g., A^2, A^1, A^0)

Chapter 3

Exponent	the power of a number
Operand	a quantity or symbol to be operated upon by an operator

Operator	a mathematical symbol representing a mathematical process
Sigma	summation
Square	any number multiplied by itself

Chapter 4

AND	Boolean expression for the concept of switches in series
Disabled	this describes a gate which is not conducting current
Enabled	this describes a gate which is conducting current
Gate	a circuit component providing an output from the circuit
Inverter	circuitry component that has the ability to change 1 to 0 and 0 to 1
Associative laws	when three or more items are added or multiplied, the order of addition or multiplication is inconsequential
Commutative laws	the order of addition or multiplication have no effect on the result
Distributive law	if two (or more) numbers are to be added, then multiplied by a third number, the result is the same if each number is first multiplied, then the products added together
Load	a component of a circuit that provides a use for electrical power
OR	Boolean expression for the concept of switches in parallel
Parallel	switches connected side-by-side

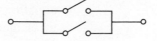

Postulate	proposition that may be taken for granted
Schematic	line drawing of a circuit or part of a circuit
Series	switches connected end-to-end

Source	source of power in a circuit
Theorem	rule that can be proven to be true
Truth tables	a method used to prove Boolean theorems

Chapter 5

Antecedent	first premise
Biconditional	statement that is true only if both conditions are true or both conditions are false
COBOL	COmmon Business Oriented Language
Compound	conditional expression containing two or more conditions
Conditional	a statement that gives one or more conditions from which to choose
Conjunction	logical AND
Consequent	second premise
Disjunction	compound logical OR
Equivalent	two or more expressions with identical meaning
Exclusive OR	one or the other of the conditions are true, but both cannot be true
Field	an area designed to hold a particular piece of information
Implied	factors may be implied instead of being repeated; both the subject and operator may be implied under the proper conditions
Simple	conditional expression containing just one condition

Chapter 6

Additive inverse	adding the inverse of a known element to the other side of the equation, to simplify the equation (this is really adding to both sides)
Constant	an element used in an equation that does not change in value
Continuous	a smooth line on a graph upon which continuous measurements may be taken
Discontinuous	specific points on a graph, where no additional measurements are possible
Domain	the first set in a function

Function	the relationship between two sets where each element of the first set is related to just one element of the second set
Image	the element of the range as it relates to the element of the domain
Like terms	two or more symbols that are alike
Mapping	showing the relationship of elements of the domain and range of the function
Multiplicative inverse	multiplying both sides of an equation with the inverse of a known element, to simplify the equation
Open sentence	algebraic statement containing one or more missing element(s)
Point of origin	the zero point on a graph
Quadrants	the four rectangles formed by a Cartesian coordinate graph
Range	elements of the second set that are assigned to elements of the first set
Replacement set	a set of numbers, each of which can replace an unknown in an equation
Root	a number that replaces a symbol which was an unknown in an equation
Sentences	statements written in symbols
Variable	an element in an equation that will have more than one value associated with it

Chapter 7

Dependent	no solution to matrix possible as there will be an infinite number of intersections
Determinant	the single value of a square matrix obtained from the sum of the products formed in accordance with a specific set of rules
Diagonal	a square matrix with all elements not on the main diagonal being zeros
Equal matrices	two matrices in which the corresponding elements are identical
Equivalent matrices	two matrices that are the same size, i.e., have the same number of rows and columns
Inconsistent	no solution to matrix possible as there will be no point of intersection

Main diagonal	in a square matrix, the elements from the upper left corner through the lower right corner
Matrix	an array of numbers of symbols, arranged in a sequence of rows and columns
Size (of a matrix)	the number of rows and columns in a matrix
Square matrix	a matrix which contains the same number of rows as columns
Submatrices	smaller sized square matrices made from a single rectangular matrix
Symmetric matrix	one in which the transpose is identical to the original matrix
Transpose	to interchange rows and columns of a matrix
Triangular matrix	a square matrix containing all zeros either above or below the main diagonal
Vector	a single row matrix or a single column matrix

E Bibliography of Related Reference Materials

Boole, George, *Studies in Logic and Probabilities*. London: Watts and Co., 1952.

Campbell, Hugh G., *Matrices, Vectors, and Linear Programming*. New York: Appleton-Century-Crofts, 1965.

Cullen, Charles G., *Matrices and Linear Transformations*. Reading, Mass.: Addison-Wesley Publishing Co., Inc., 1966.

DeAngelo, Salvatore, and Paul Jorgensen, *Mathematics for Data Processing*. New York: McGraw-Hill Book Company, 1970.

Hohn, Franz E., *Applied Boolean Algebra*. New York: The Macmillan Company, 1966.

Lewis, Laurel J., Donald K. Reynolds, F. Robert Berseth, and Frank J. Alexandro, Jr., *Linear Systems Analysis*. New York: McGraw-Hill Book Company, 1969.

Maddox, Thomas K., and Lawrence H. Davis, *Elementary Functions*. Englewood Cliffs, N. J.: Prentice-Hall, Inc., 1969.

Mayor, John R., and Marie S. Wilcox, *Contemporary Algebra*. Englewood Cliffs, N. J.: Prentice-Hall, Inc., 1965.

Morse, Anthony P., *A Theory of Sets*. New York: Academic Press, Inc., 1965.

Pettofrezzo, Anthony J., and Donald W. Hight, *Number Systems.* Glenview, Ill.: Scott, Foresman & Company, 1969.

Phister, Montgomery, Jr., *Logic Design of Digital Computers.* New York: John Wiley & Sons, Inc., 1963.

Price, Wilson T., and Merlin Miller, *Elements of Data Processing Mathematics.* New York: Holt, Rinehart & Winston, Inc., 1967.

Saxon, James A., COBOL: *A Self-Instructional Manual,* 2nd ed. Englewood Cliffs, N. J.: Prentice-Hall, Inc., 1971.

Saxon, James A., and Wesley W. Steyer, *Basic Principles of Data Processing.* 2nd edition. Englewood Cliffs, N. J.: Prentice-Hall, Inc., 1970.

Zehna, Peter W., and Robert L. Johnson, *Elements of Set Theory.* Boston: Allyn and Bacon, Inc., 1962.

INDEX

243